Two- and Three-Dimensional Flow of Groundwater

Two- and Three-Dimensional Flow of Groundwater

Florimond De Smedt and Wouter Zijl

CRC Press
Taylor & Francis Group
Boca Raton London New York

CRC Press is an imprint of the
Taylor & Francis Group, an **informa** business

CRC Press
Taylor & Francis Group
6000 Broken Sound Parkway NW, Suite 300
Boca Raton, FL 33487-2742

Printed on acid-free paper

International Standard Book Number-13: 978-1-138-57888-3 (Hardback)

Visit the Taylor & Francis Web site at
http://www.taylorandfrancis.com

and the CRC Press Web site at
http://www.crcpress.com

Contents

List of figures

Preface

This work deals with the mathematical aspects of groundwater flow and is intended to serve students and practitioners by bridging the gap between basic hydrogeology and groundwater modeling. Groundwater is the world's largest freshwater source, but sound and sustainable exploitation remains a challenge. Groundwater resources cannot be developed or managed without proper knowledge of the physical and mathematical laws governing the state and movement of groundwater. In the last decades, substantial progress has been made in groundwater modeling, and so modeling has become an essential tool for groundwater resources management and often also of environmental engineering. However, sufficient knowledge of basic mathematical principles of groundwater flow is essential to understand numerical models. Often such knowledge is lacking in traditional academic programs such as civil engineering or geology. Hence, this work attempts to synthesize the mathematics of groundwater flow and to provide information in an easily accessible form for hydrogeologists and engineers pursuing bachelor's or master's degrees as well as for groundwater practitioners and consultants who intend to become skillful and competent modelers.

Inspiration for this work came from our involvement as teacher and supervisor in the water resources engineering MSc course at Katholieke Universiteit Leuven (KU Leuven) and Vrije Universiteit Brussel (VUB), Belgium. The program, basically intended for foreign students mostly from developing countries, has been running for a couple of decades and has close to a thousand graduates working as researchers, consultants, academics, and practitioners all over the world. Students have to select a minor course that deals with groundwater, which includes specialized courses, with the ultimate goal of becoming proficient in groundwater modeling. However, students come from different backgrounds with diverse academic degrees in engineering, geological sciences, or environmental sciences. Engineers may have a profound knowledge of the physical and mathematical laws of conservation and movement but often lack insight into specific properties and settings of groundwater bearing formations and how these affect the equations describing groundwater flow. For earth scientists, it is usually the opposite as they have good knowledge about subterranean geophysical settings and processes but often lack insight into mathematical aspects for describing flow and transport. Over time, it became clear that a link was needed from basic groundwater hydrology to groundwater modeling. We believe this work provides such a link and contributes to a better understanding of groundwater flow theory and provides a greater and more realistic insight into what groundwater models can do and how they should be applied in practice.

Chapter 1 explains how the physical laws of conservation of mass and conservation of momentum are applied to groundwater flow and enable formulation of the basic laws of groundwater flow. Chapter 2 discusses different forms and simplifications of groundwater flow equations and mathematical expressions for the most common types of boundary conditions. Chapter 3 is devoted to the hydraulic approach of groundwater flow, whereby dimensions are reduced by considering flow to be restricted in certain directions depending on the geological setting of groundwater bearing formations, which is a common approach in many groundwater models. The last chapter provides some well-known analytical solutions of widely used groundwater flow problems that enable derivation of useful information about groundwater heads and flows in relation to porous medium properties. The references are not intended to be comprehensive but direct the reader to works that can be considered as milestones in the development of groundwater flow theory.

Acknowledgments

This work was developed over a period of years. An early version was published in 1999 as a chapter in the *Handbook of Groundwater Engineering* and later revised in 2007 in the second edition. The present work was first intended as an update for the third edition of the handbook, but due to some unforeseen circumstances missed the publication. Therefore, we are very grateful to CRC Press for giving us the opportunity to publish this work as a separate monograph.

About the authors

Florimond De Smedt received his PhD in civil engineering (hydrology) from Vrije Universiteit Brussel (VUB) in 1979, MSc in agronomy (soil physics) from New Mexico State University in 1979, and MSc in civil engineering from VUB in 1973. He is professor emeritus and former head of the Department of Hydrology and Hydraulic Engineering, VUB, and presently part-time professor at the same institute. He has been teaching courses for more than 30 years on hydrogeology, groundwater modeling, hydraulics, and soil mechanics. His main fields of research are groundwater modeling, water resources engineering, and physical land resources engineering. He has been supervisor of 29 PhD theses and about 100 MSc theses and is author or coauthor of about 120 journal papers with impact factor, 10 books or chapters in books, and about 200 other scientific contributions.

Wouter Zijl received his PhD in 1978 from Eindhoven University of Technology (TU/e) in technical physics, with emphasis on free surface flow applied to two-phase flow. In 1973, he received his MSc in technical physics, also from TU/e, on ionization phenomena in shock waves. In 1966, he received his BSc in electrical engineering from the Higher Institute of Technology, Amsterdam (HTS-A). He has been working as head of the Thermal Hydraulics Department at Neratoom B.V. in The Hague and as senior geophysicist, petroleum reservoir engineer, and hydrogeologist at the Netherlands Organization for Applied Scientific Research. Presently, he is visiting professor at Vrije Universiteit Brussel (VUB). His expertise includes fluid dynamics in clear and porous media, geophysical modeling (imaging, calibration, data assimilation, scale analysis), geomechanics, and applied mathematical physics in general. He is a member of the International Association of Hydrologists (IAH) and of the European Association of Geoscientists and Engineers (EAGE).

Fundamentals 1

1.1 INTRODUCTION

This chapter is intended as a guide to the laws governing groundwater flow through the porous subsurface. These laws relate the kinematics of the water "particles"—the time-dependent motion of the "particles" through three-dimensional space—to the dynamics—the forces causing groundwater flow. The fundamental principles are conservation of mass and conservation of momentum. These principles lead to mathematical equations—generally, partial differential equations—in three-dimensional space and time, in 4-D space-time, so to speak. It is important to keep the 4-D "picture" underlying the fundamental equations in mind when working on practical groundwater problems. However, the equations can often be simplified by reducing them to two- or even one-dimensional in space, sometimes also neglecting the time-dependent term.

The purpose of this chapter is to obtain flow equations that enable to cal culate the state variables of groundwater in each point of the three-dimensional flow domain and in time. Therefore, groundwater flow equations are expressed in the form of partial differential equations, with the three spatial Cartesian coordinates x, y, z and time t as independent variables. Hence, the variables that describe the state of groundwater are explicit functions of position and time. For instance, the focus of groundwater hydrology is generally directed to the determination of the hydraulic head, h, as differences in hydraulic head cause groundwater flow, which is quantified by the flux density, or briefly the flux, $\mathbf{q}$, a vector with Cartesian components q_x, q_y, and q_z. Hence, the coordinates

x, y, z, and t are independent variables, while state variables as h and $\mathbf{q}$ are considered dependent, i.e., with each set of values (x,y,z,t) variables can be associated as $h(x,y,z,t)$ and $\mathbf{q}(x,y,z,t)$.

Variables have dimensions. Whenever a new variable is introduced, the dimensions will be indicated between brackets []. The symbols used are [L] for length, [T] for time, [M] for mass, and [F = ML/T^2] for force. For practical applications, the dimensions can be substituted by any consistent set of units. A logical choice would be the International Systems of Units, i.e., meter, second, kilogram, and Newton, respectively; however, because groundwater flow is usually extremely slow, often days are used instead of seconds for the time dimension.

1.2 MASS BALANCE EQUATION

The first fundamental law governing groundwater flow is the mass balance or continuity equation. Consider in a fully saturated soil an elementary control volume centered on a point with Cartesian coordinates (x,y,z) as shown in Figure 1.1. It is customary and convenient to choose the z-axis vertical and pointing upward in the positive direction. The size of the elementary volume is Δx in the x direction, Δy in the y direction, and Δz in the z direction.

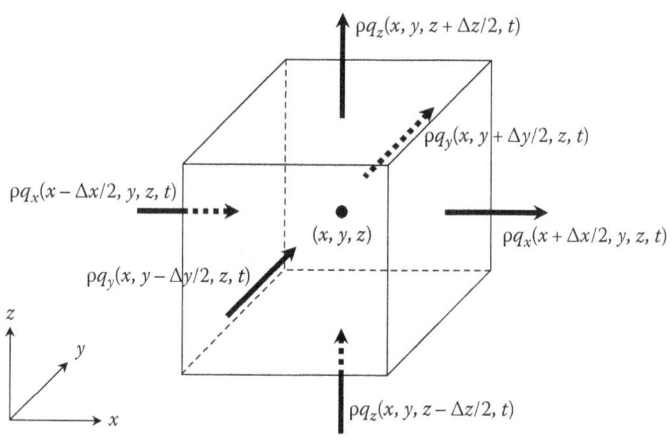

FIGURE 1.1 Mass conservation of the water phase in a reference elementary volume of saturated porous material.

At a certain time instant, t, the mass of groundwater, m, present in the elementary control volume is given by

$$m = \rho n \Delta x \Delta y \Delta z, \tag{1.1}$$

where

ρ is the density of water with dimensions [M/L^3]
n is the porosity of the porous medium with dimensions [L^3/L^3]

The principle of mass conservation implies that the net result of inflow minus outflow is balanced by the change in storage versus time, or

$$\frac{\partial m}{\partial t} = \text{Inflow} - \text{Outflow}. \tag{1.2}$$

Hence, it is necessary to calculate the groundwater flows through the sides of the elementary control volume in order to evaluate the net result of inflow minus outflow. The amount of groundwater flow is denoted by means of the flux, $\mathbf{q} = (q_x, q_y, q_z)$, with dimensions [$L^3/T/L^2 = L/T$], which is the volumetric discharge or flow rate per cross-sectional area. For instance, the mass inflow of groundwater through the left side of the control volume is given by $\rho q_x\left(x - \frac{1}{2}\Delta x, y, z, t\right)\Delta y \Delta z$. Because Δx is small, ρq_x at position $x - \frac{1}{2}\Delta x$ can be approximated by a Taylor series expansion at position x, maintaining only terms of order zero and first order, so that this inflow can be calculated as

$$\rho q_x\left(x - \tfrac{1}{2}\Delta x, y, z, t\right)\Delta y \Delta z \approx \left(\rho q_x - \tfrac{1}{2}\Delta x \frac{\partial \rho q_x}{\partial x}\right)\Delta y \Delta z, \tag{1.3}$$

where ρq_x and its derivative with respect to x are evaluated at the center of the control volume. Similar expressions can be established for the other sides; for instance, at the right side of the control volume, at position $x + \frac{1}{2}\Delta x$, the groundwater outflow is given by

$$\rho q_x\left(x + \tfrac{1}{2}\Delta x, y, z, t\right)\Delta y \Delta z \approx \left(\rho q_x + \tfrac{1}{2}\Delta x \frac{\partial \rho q_x}{\partial x}\right)\Delta y \Delta z. \tag{1.4}$$

Hence, the total inflow minus outflow can be calculated as

$$\begin{aligned}\frac{\partial m}{\partial t} &= \left(\rho q_x - \tfrac{1}{2}\Delta x \frac{\partial \rho q_x}{\partial x}\right)\Delta y \Delta z + \left(\rho q_y - \tfrac{1}{2}\Delta y \frac{\partial \rho q_y}{\partial y}\right)\Delta x \Delta z \\ &+ \left(\rho q_z - \tfrac{1}{2}\Delta z \frac{\partial \rho q_z}{\partial z}\right)\Delta x \Delta y - \left(\rho q_x + \tfrac{1}{2}\Delta x \frac{\partial \rho q_x}{\partial x}\right)\Delta y \Delta z \\ &- \left(\rho q_y + \tfrac{1}{2}\Delta y \frac{\partial \rho q_y}{\partial y}\right)\Delta x \Delta z - \left(\rho q_z + \tfrac{1}{2}\Delta z \frac{\partial \rho q_z}{\partial z}\right)\Delta x \Delta y. \qquad (1.5)\end{aligned}$$

Working out term by term and combining yields the following expression:

$$\frac{\partial m}{\partial t} = -\left(\frac{\partial \rho q_x}{\partial x} + \frac{\partial \rho q_y}{\partial y} + \frac{\partial \rho q_z}{\partial z}\right)\Delta x \Delta y \Delta z. \qquad (1.6)$$

Using the del operator $\nabla = (\partial_x, \partial_y, \partial_z) = (\partial/\partial x, \partial/\partial y, \partial/\partial z)$ [L^{-1}], this can be written as

$$\frac{\partial m}{\partial t} = -\nabla \cdot (\rho \mathbf{q})\, \Delta x \Delta y \Delta z \qquad (1.7)$$

where the dot represents the scalar product of two vectors.

Using Equation 1.1, the change in storage is also given by

$$\frac{\partial m}{\partial t} = \frac{\partial \rho n \Delta x \Delta y \Delta z}{\partial t}. \qquad (1.8)$$

In this expression, the variables that can change with time are the density of the water, ρ, because water is compressible; the porosity, n; and size of the control volume, $\Delta x \Delta y \Delta z$, because the porous medium can be compressible. However, for the latter, it is usually assumed that under normal field conditions only vertical deformation needs to be considered, so that only n and Δz depend upon time while $\Delta x \Delta y$ remains constant. Hence, Equation 1.8 can be worked out by using the rules of differentiation as

$$\begin{aligned}\frac{\partial m}{\partial t} &= \rho n \Delta x \Delta y \frac{\partial \Delta z}{\partial t} + \rho \frac{\partial n}{\partial t}\Delta x \Delta y \Delta z + \frac{\partial \rho}{\partial t} n \Delta x \Delta y \Delta z \\ &= \rho\left(\frac{n}{\Delta z}\frac{\partial \Delta z}{\partial t} + \frac{\partial n}{\partial t} + \frac{n}{\rho}\frac{\partial \rho}{\partial t}\right)\Delta x \Delta y \Delta z. \qquad (1.9)\end{aligned}$$

In addition, it is generally assumed that Δz changes only because of changes in porosity, while the solid phase may be considered as incompressible. Hence, the solid volume per unit horizontal surface area, $(1-n)\Delta z$, remains constant in time, or

$$\frac{\partial(1-n)\Delta z}{\partial t} = (1-n)\frac{\partial \Delta z}{\partial t} - \Delta z \frac{\partial n}{\partial t} = 0, \tag{1.10}$$

from which we obtain the relationship between changes in n and Δz

$$\frac{\partial n}{\partial t} = \left(\frac{1-n}{\Delta z}\right)\frac{\partial \Delta z}{\partial t}, \tag{1.11}$$

so that Equation 1.9 becomes

$$\frac{\partial m}{\partial t} = \rho\left(\frac{1}{\Delta z}\frac{\partial \Delta z}{\partial t} + \frac{n}{\rho}\frac{\partial \rho}{\partial t}\right)\Delta x \Delta y \Delta z. \tag{1.12}$$

According to Terzaghi's theory of one-dimensional consolidation, the pore space deformation, Δz, is caused by the intergranular stress (see, e.g., Bear, 1988, p. 54), which is equal to the effective stress, σ_e [F/L^2], given by

$$\sigma_e = \sigma - p, \tag{1.13}$$

where

σ [F/L^2] is the total stress caused by the weight of the overburden
p [F/L^2] is the groundwater pressure

For small deformations, the relative change in Δz is proportional to the change in effective stress, i.e.

$$\frac{1}{\Delta z}\frac{\partial \Delta z}{\partial t} = -\alpha \frac{\partial \sigma_e}{\partial t}, \tag{1.14}$$

where α [L^2/F] is the elastic compressibility coefficient of the porous formation. In groundwater hydraulics, it is generally assumed that the weight of the overburden does not change, $d\sigma = 0$ or $d\sigma_e = -dp$, so that Equation 1.14 can be written as

$$\frac{1}{\Delta z}\frac{\partial \Delta z}{\partial t} = \alpha \frac{\partial p}{\partial t}. \tag{1.15}$$

The compressibility of the water can be expressed by a similar law as

$$\frac{1}{\rho}\frac{\partial \rho}{\partial t} = \beta \frac{\partial p}{\partial t}, \tag{1.16}$$

with β [L²/F] as the compressibility coefficient of water. It is assumed that other effects on the density, as solutes or temperature, are of minor importance and can be ignored. Substitution of these relations into Equation 1.12 gives

$$\frac{\partial m}{\partial t} = \rho\left[(\alpha + n\beta)\frac{\partial p}{\partial t}\right]\Delta x \Delta y \Delta z. \tag{1.17}$$

In the derivation of Equation 1.16, we have followed the classical approach initiated by Jacob (1940) and finalized by Jacob (1950), where the elemental volume is allowed to deform. This was later criticized by, e.g., De Wiest (1966) and Cooper (1966), arguing that a control volume should not deform. A more rigorous development using a fixed control volume and fixed coordinates is possible. In this case, the movement of the solid phase has to be taken into account and also its conservation of mass. This has been achieved by, e.g., Gambolati and Freeze (1973), Freeze and Cherry (1979), and Dagan (1989), resulting in an expression that is independent of the coordinate system and that corresponds to Equation 1.17 for the Cartesian coordinate system used here. Another more general approach considers three-dimensional deformation as proposed by Biot (1955); a good overview of this approach is given by Verruijt (1969) (also see Bear, 1988, pp. 208–211).

The continuity equation is obtained by putting the change in storage, Equation 1.17, equal to the net inflow given by Equation 1.7. Dividing by $\rho\Delta x\Delta y\Delta z$ to express the mass balance per unit volume of porous medium and neglecting the small term $(\nabla\rho)/\rho$ result in

$$(\alpha + n\beta)\frac{\partial p}{\partial t} = -\nabla \cdot \mathbf{q}. \tag{1.18}$$

This relationship states the principle of mass conservation of groundwater in its most general form in a notation that is independent from the choice of a particular coordinate system. Equation 1.18 gives the change in volume of groundwater present in the porous medium due to compression of the medium and the water. The right-hand side of the equation gives the convergence or divergence of the volumetric flow rate of groundwater. When the flow is converging, the storage increases, and when it diverges the storage decreases.

1.3 MACROSCOPIC APPROACH

The obtained continuity equation is elementary, yet fundamental, but not without ambiguity. The problem arises from the fact whether Δx, Δy, and Δz can be made small enough to justify the truncation of the Taylor series to terms of orders zero and one in the derivation of the inflows and outflows, as in Equation 1.5. Indeed, when the limit is taken of Δx, Δy, and Δz going to zero, the elementary control volume reduces to a point, so that the concept of a continuous porous medium becomes illusive, because a point can be situated in the water phase or the solid phase, where state variables such as hydraulic head, groundwater pressure, and groundwater flux do not exist.

Hence, it seems that groundwater flow problems cannot be formulated and solved correctly at the microscopic level. Of course, such a microscopic approach is not really of any interest in practice. But nevertheless, this ambiguity needs to be cleared up. This is achieved by abandoning the microscopic pore-scale level and by moving to a coarser macroscopic level. At this level, microscopic features such as solid grains and in between pore spaces are ignored and the medium is conceived as a continuous space with average properties, so that porosity exists in any point of the medium regardless of whether this point is situated in the solid phase or in the pore space. Furthermore, the variables describing the state and movement of water through the medium become macroscopic, so that the average behavior is described and not the fate of individual fluid particles moving in the pores of the porous medium.

The question remains how macroscopic variables need to be defined and interpreted. Different approaches are in use. The first and still most popular technique is the method of the representative elementary volume (REV), developed by Bear (1972), where macroscopic variables are defined as volumetric averages over an REV and attributed to the center of the REV. The exact size of the REV is not determined, but it is assumed that the size is much larger than the pore scale and much smaller than the scale of the porous medium. The results obtained with this approach yield macroscopic values for the groundwater flow, representing the average behavior of the fluid over the REV. The main advantage of the method is that macroscopic quantities have a clear physical meaning, as they are measurable in the field by taking and analyzing soil samples of an adequate size. The main disadvantages are the loss of detailed information on the microscopic pore-scale level and the uncertainty due to the assumption that the values of the macroscopic variables are independent of the size of the REV. A detailed discussion can be found in Pinder (1983).

A second technique is a statistical approach taking into account the uncertainty of the spatial distribution on a microscopic scale (Dagan, 1989).

The microscopic arrangement of the porous medium is considered to be random, such that a set of media can be imagined with similar characteristics. The macroscopic variables are considered to be the ensemble mean values, or expectation values of the variables of the media in the set. The main conceptual difficulty stems from the fact that the mean value of an ensemble of realizations is not generally the same as the volumetric average over an REV. Ensemble mean value and volumetric average are equal only under the assumption of ergodicity, i.e., under the assumption that the characteristics of one sample are assumed to be representative for the whole set. With the theory of random functions, macroscopic laws can be derived. As long as only relationships between expectation values of quantities are derived, with no special concern about their fluctuations, the results obtained are essentially the same as obtained with the volume averaging REV approach. However, the statistical technique is more powerful when dealing with fluctuations and deviations, as in case of particle or solute transport in groundwater.

1.4 EQUATION OF MOTION

The second fundamental law is the momentum balance, which is based on Newton's second law of motion, i.e., forces induce motion or a change in motion. Consider the elementary control volume as used previously when deriving the continuity equation. An inventory of all forces acting on the water present in the control volume can be made in order to obtain a momentum equation. Under natural conditions, the forces to consider are pressure forces, gravity force, and reaction forces of the solid matrix exerted on the fluid. As forces are vectors, it is necessary to consider different components along the different directions, as shown in Figure 1.2. For instance, along the left side of the control volume, pressure is acting on the water phase, which yields $np\left(x-\frac{1}{2}\Delta x,y,z,t\right)\Delta y\Delta z$ to the force balance in the x direction. The porosity, n, appears in the expression because the water occupies only that fraction of the boundary. A similar force is acting on the right side, but in opposite direction, i.e., $-np\left(x+\frac{1}{2}\Delta x,y,z,t\right)\Delta y\Delta z$. Using truncated Taylor series expansions to relate the np-terms to the center of the control volume, the resulting pressure force component in the x direction becomes

$$\left[\left(np-\tfrac{1}{2}\Delta x\frac{\partial np}{\partial x}\right)-\left(np+\tfrac{1}{2}\Delta x\frac{\partial np}{\partial x}\right)\right]\Delta y\Delta z=-\frac{\partial np}{\partial x}\Delta x\Delta y\Delta z. \quad (1.19)$$

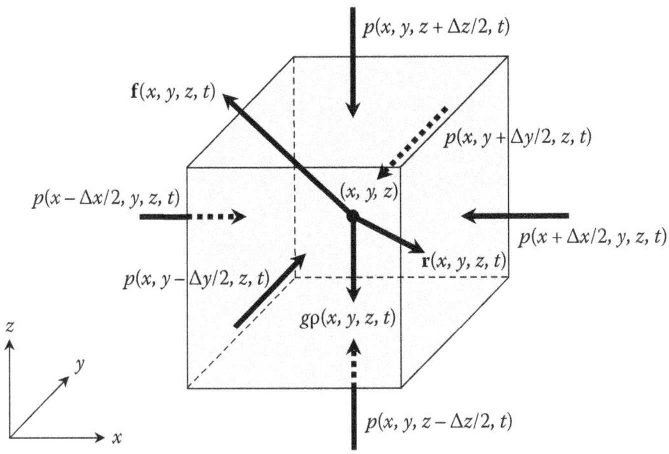

FIGURE 1.2 Forces acting in the water phase of a reference elementary volume of saturated porous material.

Similar expressions can be obtained for the pressure force components in the y and z directions, acting on the other sides of the elementary control volume. The gravity force is the total weight of the water in the control volume; hence, $-\rho g n \Delta x \Delta y \Delta z$ with g as the gravity constant [L/T^2], acting downward in the z direction.

The evaluation of the reaction force of the solid material on the water is more complicated. It consists of forces acting against the water pressure and friction forces due to the groundwater movement. These forces are extremely difficult to evaluate on a pore-scale level, because the shape of the contact surface between the solid phase and the water phase is very complex from a geometrical point of view. Because the exact contact surface is generally not known and would be very difficult to express in mathematical terms anyway, it is impossible to describe these forces on a microscopic scale, and one is forced to adopt a less precise macroscopic approach. Therefore, the reaction forces are defined as average body forces per water volume; the reaction force against the water pressure is denoted as $\mathbf{r}=(r_x, r_y, r_z)$ [F/L^3], and the friction force against water movement as $\mathbf{f}=(f_x, f_y, f_z)$ [F/L^3]. Hence, the effect of these forces in the x direction is $(r_x+f_x)n\Delta x\Delta y\Delta z$, and similar expressions can be derived in the y and z directions.

Dividing by the total volume of water present in the control volume, $n\Delta x\Delta y\Delta z$, the force balance per unit volume in the three directions can be calculated, respectively, as

$$r_x + f_x - \frac{1}{n}\frac{\partial np}{\partial x}, \quad r_y + f_y - \frac{1}{n}\frac{\partial np}{\partial y}, \quad r_z + f_z - \rho g - \frac{1}{n}\frac{\partial np}{\partial z}. \tag{1.20}$$

Using the del operator, ∇, these can be combined in one vector equation

$$\mathbf{r} + \mathbf{f} - \rho g \nabla z - \nabla p - \frac{p}{n} \nabla n, \tag{1.21}$$

(where z is assumed to be positive in upward direction). Next, note that if the fluid is at rest, the sum of all forces should be zero. Also, there would be no friction, $\mathbf{f}=0$, and the pressure should be hydrostatic, $\nabla p = -\rho g \nabla z$, from which the overall reaction force, $\mathbf{r}$, of the grains opposing the water pressure can be evaluated as (Dagan, 1989, p. 73)

$$\mathbf{r} = \left(\frac{p}{n}\right) \nabla n. \tag{1.22}$$

Substituting this result in the force balance simplifies Equation 1.21 to

$$-\nabla p - \rho g \nabla z + \mathbf{f}. \tag{1.23}$$

In case of motion, the sum of forces is not zero but equals the change in momentum of the fluid. Furthermore, the friction along the solid–water interface is nonzero and should be specified as a function of the flow and porous medium properties. Several additional assumptions and considerations are necessary to arrive at a useful result.

From field observations, it is known that groundwater flow under natural conditions is generally very slow, which leads to important simplifications. First, changes in momentum are also very small and can be neglected compared to other forces acting on the fluid. Hence, although the fluid is in motion, the forces acting on the fluid may be assumed in equilibrium

$$-\nabla p - \rho g \nabla z + \mathbf{f} = 0. \tag{1.24}$$

This type of flow is known in fluid dynamics as creeping motion. The active forces yielding motion, such as pressure and gravity, are immediately balanced by resisting friction forces of equal strength. Of course, in porous media, this is due to the large contact area between fluid and solid material, which causes extensive friction, such that significant movement of the fluid is prevented.

Second, because water is a viscous fluid, the friction force results from viscous momentum transfer between the stagnant pore walls and the moving fluid. In principle, the Navier–Stokes equations should enable to calculate the overall resistance force, but due to the complex geometry of the pore space this is unfeasible in practice. However, in case of viscous creeping flow, it is known that the overall resistance force is opposite in direction to the fluid flow, proportional to the viscosity of the fluid and the magnitude of the flow, and depending on the size of the obstacles in the flow field. Hence, it is justified to express the friction force as

$$\mathbf{f} = -\left(\frac{\mu}{k}\right)\mathbf{q}, \tag{1.25}$$

where μ [FT/L^2] is the dynamic viscosity of the fluid, $\mathbf{q}$ is the flux as defined in the previous paragraph, here representing the amount and direction of the water movement, and k [L^2] is denoted as absolute or intrinsic permeability, or sometimes permeability in short, which represents the geometry of the pore space.

Substitution of Equation 1.25 into the force balance Equation 1.24 and rearranging result in the following equation of motion:

$$\mathbf{q} = -\left(\frac{k}{\mu}\right)\left(\nabla p + \rho g \nabla z\right). \tag{1.26}$$

The obtained expression, called Darcy's law, describes the flow of a fluid in a porous medium, also if the fluid has a variable density and viscosity. Such conditions can be present in, for instance, coastal aquifers where salt and fresh water intermix, or in geothermal reservoirs where the density and viscosity of the fluid change with temperature.

When the fluid density is constant, Darcy's law can be simplified as

$$\mathbf{q} = -\left(\frac{\rho g k}{\mu}\right)\nabla\left(\phi + z\right) = -K\nabla h, \tag{1.27}$$

where

$K = \rho g k/\mu$ [L/T] is the hydraulic conductivity

$\phi = p/\rho g$ is the pressure head [L]

$h = \phi + z$ is the hydraulic head, or briefly, the head

Hence, Darcy's law, which originally was based on experimental findings, is nothing else but an expression of equilibrium of forces in case of viscous creeping flow in porous media. The derivation presented here sheds more light on the underlying principles and assumptions that result in Darcy's law and enables to appreciate more its applicability in field conditions.

1.5 DARCY'S LAW

Darcy's law is a linear vector relationship between flux vector, $\mathbf{q}$, and head gradient vector, ∇h, i.e.

$$\mathbf{q} = -K\nabla h, \tag{1.28}$$

which in Cartesian coordinates with flux vector components (q_x, q_y, q_z) and head gradient vector components $(\partial h/\partial x, \partial h/\partial y, \partial h/\partial z) = (\partial_x h, \partial_y h, \partial_z h)$ becomes

$$q_x = -K\frac{\partial h}{\partial x}, \quad q_y = -K\frac{\partial h}{\partial y}, \quad q_z = -K\frac{\partial h}{\partial z}. \tag{1.29}$$

The basic assumption leading to Darcy's law is that movement of a fluid through a porous medium is very restricted, due to large friction forces that balance the driving forces for motion. Hence, Darcy's law is applicable in cases as water flow in soils or other types of granular porous media, or flow in fractured rocks, but not in cases as caves or other large-sized openings like cracks, fissures, etc. These latter problems should be analyzed with viscous flow theories. Second, all ambiguity as discussed in the case of the mass balance equation is also present here. Darcy's law is a macroscopic approach; on a microscopic level, there are no such things as hydraulic conductivity or permeability. Hence, one might wonder about the errors involved in using a macroscopic approach as Darcy's law. However, this is not really an important issue, because the uncertainty in obtaining representative and accurate values for the hydraulic conductivity or permeability has a much larger effect on the overall accuracy. Indeed, field investigations show that natural porous media such as aquifers and aquitards exhibit a large variability in conductive properties, which is difficult to quantify accurately by experimental or deterministic means.

In the Darcy's law presented above, we have assumed that the porous medium is isotropic, so that the hydraulic conductivity does not depend upon direction. However, in case of direction-related properties such as preferential lining of fractures, stratifications or layering, the hydraulic conductivity can depend upon direction. In such cases, the relationship between flux vector, $\mathbf{q}$, and head gradient vector, ∇h, becomes

$$\mathbf{q} = -\underline{\underline{\mathbf{K}}} \cdot \nabla h, \tag{1.30}$$

where the hydraulic conductivity is now represented by a second-order tensor, $\underline{\underline{\mathbf{K}}}$. It can also be proved that, on the scale of a REV, this tensor is symmetric

(Olmstead, 1968). For coarser scales, it depends on the way of upscaling whether it is symmetric or not (Zijl and Nawalany, 1993). In Cartesian coordinates, the hydraulic conductivity tensor becomes a 3 by 3 matrix with components K_{xx}, K_{xy}, K_{xz}, etc., so that Darcy's law is written as

$$\begin{bmatrix} q_x \\ q_y \\ q_z \end{bmatrix} = -\begin{bmatrix} K_{xx} & K_{xy} & K_{xz} \\ K_{yx} & K_{yy} & K_{yz} \\ K_{zx} & K_{zy} & K_{zz} \end{bmatrix} \cdot \begin{bmatrix} \partial_x h \\ \partial_y h \\ \partial_z h \end{bmatrix} = \begin{bmatrix} -K_{xx}\partial_x h - K_{xy}\partial_y h - K_{xz}\partial_z h \\ -K_{yx}\partial_x h - K_{yy}\partial_y h - K_{yz}\partial_z h \\ -K_{zx}\partial_x h - K_{zy}\partial_y h - K_{zz}\partial_z h \end{bmatrix}. \quad (1.31)$$

Hence, under conditions of anisotropy, a head gradient in one direction can yield flows in other directions. Especially in fractured porous media, anisotropy can be important. However, practical applications remain scarce because the components of the hydraulic conductivity tensor are difficult to determine in the field.

Anisotropy also plays an important role in layered formations, usually of sedimentary origin. For horizontal layering, only two conductivity components exist: a horizontal conductivity, K_h, and a vertical conductivity, K_v. In such a case, Darcy's law becomes

$$q_x = -K_h \frac{\partial h}{\partial x}, \quad q_y = -K_h \frac{\partial h}{\partial y}, \quad q_z = -K_v \frac{\partial h}{\partial z}. \quad (1.32)$$

These equations are useful in practice, also in view of the fact that effects of horizontal layering on the conductivity are rather the rule than the exception in ground layers. In the case of folded or dipping layered formations also the off-diagonal terms are important. Following the rules of tensor calculus (Bear, 1972), all components of the conductivity tensor can be calculated from the dip angle and the conductivity values normal and parallel to the layering. In this context, it is important to mention that, in addition to anisotropy, heterogeneity, i.e., conductivity variations from point to point, is also rather the rule than the exception and often complicates practical applications.

Groundwater Flow Equations

2

2.1 TRANSIENT GROUNDWATER FLOW

The equation for transient groundwater flow is obtained by combining the mass conservation equation with Darcy's law. The most general form is obtained when Equation 1.18 is combined with the general Darcy equation (1.26), yielding

$$(\alpha + n\beta)\frac{\partial p}{\partial t} = \nabla \cdot \left[\left(\frac{k}{\mu}\right)(\nabla p + \rho g \nabla z)\right]. \tag{2.1}$$

Note that the porosity, n, on the left-hand side of the equation, depends upon the water pressure due to elastic deformation of the porous medium; actually, the same holds for the absolute permeability, k. However, these effects are generally ignored and the groundwater flow equation is expressed as a linear partial differential equation for the water pressure as follows:

$$S_p \frac{\partial p}{\partial t} = \nabla \cdot \left[\left(\frac{k}{\mu}\right)(\nabla p + \rho g \nabla z)\right], \tag{2.2}$$

where $S_p = \alpha + n\beta$ [L^2/F] is the storage coefficient of the porous medium related to water pressure changes. This storage coefficient depends upon the compressibility of the porous medium and the fluid but has a simple physical interpretation, i.e., the storage coefficient is the volume of water released by a unit volume of the porous medium due to a unit decline of the water pressure. Equation 2.2 describes three-dimensional movement of groundwater when density effects have to be taken into account. Notice that, as a consequence, the flow equation is expressed as a function of the water pressure, and not as a function of the head. A state of the art of density-driven flow in porous media is, for instance, given by Holzbecher (1998).

When density-driven flow can be ignored, a groundwater flow equation can be obtained by combining the continuity equation with Darcy's law given by Equation 1.28, which yields

$$(\alpha + n\beta)\frac{\partial p}{\partial t} = \nabla \cdot (K \nabla h). \tag{2.3}$$

Making use of the fact that when the fluid density is constant the water pressure differences in time can be related to the temporal variation of the head, i.e., $\partial p / \partial t = \rho g \partial h / \partial t$, the resulting flow equation can be written as

$$S_s \frac{\partial h}{\partial t} = \nabla \cdot (K \nabla h), \tag{2.4}$$

where $S_s = \rho g(\alpha + n\beta)$ $[L^{-1}]$ is the specific storage coefficient, which can be interpreted physically as the volume of water released by a unit volume of saturated porous medium due to a unit decline of the head. Common values are usually very small because elastic deformation of ground layers or water is limited and mostly not noticeable in the field. In this approach, plastic deformation of ground layers, consisting of materials like clay or peat, is excluded. Such deformations mainly occur under the influence of external loads or extensive groundwater pumping and are noticeable in the field as land subsidence. Elastic deformation and storage in aquifers are described by Verruijt (1969). Theories on porous medium deformability under various soil conditions, with special attention to the subject of land subsidence, are discussed in part 2 of Bear and Corapcioglu (1984).

Equation 2.4 can be considered as a basic groundwater flow equation, because it enables to calculate the head, h, in function of characteristics of the porous medium and the fluid. Written in Cartesian coordinates, the groundwater flow equation becomes

$$S_s \frac{\partial h}{\partial t} = \frac{\partial}{\partial x}\left(K \frac{\partial h}{\partial x}\right) + \frac{\partial}{\partial y}\left(K \frac{\partial h}{\partial y}\right) + \frac{\partial}{\partial z}\left(K \frac{\partial h}{\partial z}\right). \tag{2.5}$$

This is the most used groundwater flow equation in practice, although different hydraulic conductivities are often considered in horizontal and vertical directions. The equation is of the diffusive type, indicating that head differences will be dissipated through the medium with a speed depending upon the value of K/S_s. Hence, because S_s is usually small, this process can be rather fast, unless K is also small as for instance in case of clay layers.

It is also instructive to write the groundwater flow equation in cylindrical coordinates (r, θ, z)

$$S_s \frac{\partial h}{\partial t} = \frac{1}{r} \frac{\partial}{\partial r}\left(rK \frac{\partial h}{\partial r} \right) + \frac{1}{r^2} \frac{\partial}{\partial \theta}\left(K \frac{\partial h}{\partial \theta} \right) + \frac{\partial}{\partial z}\left(K \frac{\partial h}{\partial z} \right). \tag{2.6}$$

This equation is used for the prediction of groundwater flow toward wells, as for instance described by Hantush (1964), Walton (1970), Neuman (1972), Huisman (1972), and Kruseman et al. (1990).

2.2 STEADY-STATE GROUNDWATER FLOW

When groundwater flow is stationary, the state variables become independent of time and the flow equation, Equation 2.4, is reduced to the Laplace-type equation

$$\nabla \cdot (K \nabla h) = 0. \tag{2.7}$$

The Laplace-type equation, also called the potential equation, plays an important role in many branches of science and technology ranging from biomedical imaging to nondestructive testing and geo-electrical sounding (also see Section 2.5).

Furthermore, when regarding flow on longer timescales (weeks and months) in the shallow parts of the subsurface, the term $S_s \partial h/\partial t$ accounting for water compression and pore space deformation is negligible and Equation 2.4 reduces to Equation 2.7. However, under conditions where the flow may be considered as incompressible, i.e., in cases where Equation 2.7 is a good approximation, the flow is not necessarily steady. The boundary conditions (Section 2.3) may be time-dependent, which makes the groundwater flow time-dependent too. Especially the continuously rising and falling of the water table is a main cause of time-dependency. The time-dependency of the moving water table is described in Section 2.4.

Written in Cartesian coordinates, and generalizing by assuming homogeneous hydraulic conductivity, Equation 2.7 becomes

$$\frac{\partial^2 h}{\partial x^2}+\frac{\partial^2 h}{\partial y^2}+\frac{\partial^2 h}{\partial z^2}=0. \tag{2.8}$$

This equation clearly shows that movement of groundwater is a potential flow problem, in which the head is the potential. Many solutions have been obtained for groundwater flow problems based on the theory of harmonic functions, i.e., functions that honor potential equation (2.8), e.g., Polubarinova-Kochina (1962), Harr (1962), Verruijt (1970), Halek and Svec (1979), and Strack (1989).

Another approach, when dealing with so-called incompressible flow, is to write Equation 2.7 as the continuity equation using the flux vector

$$\nabla\cdot\mathbf{q}=\frac{\partial q_x}{\partial x}+\frac{\partial q_y}{\partial y}+\frac{\partial q_z}{\partial z}=0. \tag{2.9}$$

This equation states that the flow is divergence-free, or solenoidal. Hence, it is possible to introduce a vector potential, $\boldsymbol{\psi}$, in such a way that the flux equals the curl (or rotational) of the vector potential, i.e., $\mathbf{q}=-\nabla\times\boldsymbol{\psi}$. In this way, Equation 2.9 is honored automatically, because $\nabla\cdot(\nabla\times\boldsymbol{\psi})=0$. For more details, see Mohammed et al. (2009). In two-dimensional flow problems, the vector potential simplifies to a scalar function, the so-called stream function, ψ, such that for homogeneous and isotropic conditions the flux components can be expressed as

$$q_x=\frac{\partial\varphi}{\partial x}=\frac{\partial\psi}{\partial y},\quad q_y=\frac{\partial\varphi}{\partial y}=-\frac{\partial\psi}{\partial x}, \tag{2.10}$$

with $\varphi=-Kh$. Equation 2.10 shows that the potential, φ, and stream function, ψ, fulfill the Cauchy–Riemann conditions, which enables to formulate Darcy's law by means of complex variables, as

$$\omega=\frac{d\varsigma}{dz}, \tag{2.11}$$

where

$\omega=q_x-iq_y$ is the conjugate complex flux
$\varsigma=\varphi+i\psi$ is the complex potential
$z=x+iy$ is the complex space coordinate

Hence, solutions to two-dimensional groundwater problems can be obtained by finding relationships between these complex variables, in particular by conformal mapping, e.g., Polubarinova-Kochina (1962), Harr (1962), and Strack (1989).

2.3 INITIAL AND BOUNDARY CONDITIONS

Groundwater flow equations are partial differential equations. When considering forward problems, i.e., problems in which the parameters—hydraulic conductivity and specific storage—are specified, such equations relate the dependent variables head and flux to the independent variables, the coordinates of the flow domain and the time. However, flow equations are only valid inside the flow domain, and not on the boundaries, because there the porous medium ends and other phenomena occur, which will influence the groundwater flow inside the domain. These interactions at the boundaries are governed by other physical laws and have to be described separately by mathematical expressions, the so-called boundary conditions.

First, consider the time dimension. Groundwater flow equations in which specific storage plays a role contain the first-order partial time derivative $S_s \partial h/\partial t$ (see Equation 2.4). For such a diffusion-type equation, it can be shown mathematically that only the condition at the start, i.e., at time zero, is needed. Hence, knowing the initial distribution of the dependent variable in the flow domain is sufficient to describe the further evolution in time. For instance, if the head is the dependent variable, the so-called initial condition is

$$h(x,y,z,0) = h_0(x,y,z), \tag{2.12}$$

where h_0 represents a known function of x, y, and z.

It is generally a cumbersome task to define realistic initial conditions in practical field applications. Fortunately, in the shallower parts of the subsurface, where the effect of the specific storage term in Equation 2.4 is generally small, the steady-state equation (2.7) is a good approximation and can be used to determine realistic and physically acceptable initial conditions.

Boundary conditions for the physical domain are somewhat more complicated. Mathematically, it can be shown that, because groundwater flow equations contain second-order partial derivatives with respect to the space coordinates, conditions are required at every point of the closed boundary of the modeling domain, even when the boundaries tend to infinity. Because the interaction between the inside groundwater flow and the outside world can be complex, and our ability to conceive such phenomena accurately is rather limited, boundary conditions are usually described in a simplified way.

Under the assumption that the hydraulic conductivity is known everywhere in the modeling domain, only one boundary condition must be specified in each point on the boundary and in each well (a well may be considered as in internal boundary). Three types of boundary conditions are generally considered. First-type boundary conditions (also called Dirichlet boundary conditions) apply when the value of the dependent variable at the boundary is known. For instance, in case of heads, such a boundary condition would be

$$h\left(x_b, y_b, z_b, t\right) = h_b\left(t\right), \tag{2.13}$$

where

(x_b, y_b, z_b) represents a point on the boundary
h_b is a known function of time

This expression may be called a head boundary condition and is used when the groundwater is in contact with a water body having a known head, h_b, as for instance, a river, lake, or reservoir. Furthermore, a head measured in a well or in a piezometer may be specified as an internal head boundary condition. Of course, doing so the flux into or out of that well may not be specified. The key assumption is that whatever groundwater flow occurs at the inside of the flow domain and at the boundary, it will have no influence on the head of the outside water body, such that this head remains fixed as stated by the boundary condition.

Second-type boundary conditions (also called Neumann boundary conditions) are flux boundary conditions, where it is assumed that the amount of groundwater exchange through the boundary is known. The amount of water exchange is given by the groundwater flux component perpendicular to the boundary, so that the boundary condition can be expressed as

$$q_n\left(x_b, y_b, z_b, t\right) = q_b\left(t\right), \tag{2.14}$$

where

q_n represents the flux component normal to the boundary
$q_b(t)$ is a known function of time

A flux-type condition can also be expressed as a function of the dependent variable, the head, by using Darcy's law

$$q_n = -K \frac{\partial h}{\partial n}\bigg|_{(x_b, y_b, z_b, t)} = q_b(t), \tag{2.15}$$

where $\partial h/\partial n$ represents the derivative of the head perpendicular to the boundary. From a mathematical point of view, the choice of the positive direction of the normal to the boundary determines whether q_n values are positive or negative. However, in practice, often a simpler convention is used, by considering fluxes entering into the flow domain as positive, and outgoing fluxes as negative, regardless of the sense of the normal on the boundary.

A flux boundary condition implies that whatever the state and flow of the groundwater inside the flow domain and at the boundary, the normal flux is fixed by external conditions and remains as stated by the boundary condition. Typical examples of such conditions are pumping wells (a well may be considered as an internal boundary), groundwater recharge, infiltration, etc.; in general, any situation where the flux is supposedly fixed and known. An obvious example is an impervious boundary, as the flux component perpendicular to the boundary is strictly zero.

A third-type boundary condition (also called a Robin boundary condition) is a mixture of the two previous types. It applies when head and normal flux component at the boundary are related to each other. Such a condition applies in case of a flow domain in contact with a water body with a fixed head, h_b, but where the exchange between the groundwater reservoir and the water body is restricted due to the presence of some resistance. Mathematically, this condition is expressed as

$$q_n(x_b, y_b, z_b, t) = C_b\left[h_b(t) - h(x_b, y_b, z_b, t)\right], \tag{2.16}$$

where we have used the above mentioned sign convention for the normal boundary flux, and C_b [T^{-1}] represents the boundary conductance between the groundwater reservoir and the outside water source or sink. This parameter can be explained physically as the conductive capacity of a permeable boundary layer, present between the groundwater flow domain and the water body, such that $C_b = K_b/D_b$, where K_b is the hydraulic conductivity of the boundary layer and D_b [L] its thickness. For instance, this situation would apply when a river is in contact with the groundwater, but the interaction is restricted due to the presence of a mud layer in the riverbed. Another example is a well with a clogged filter.

2.4 WATER TABLE CONDITIONS

Actually, the water table, also often denoted as phreatic surface, is not a true boundary. For the water, there is no boundary, because it is continually flowing between the unsaturated and saturated zones. However, because the equations of unsaturated flow are usually difficult to solve, the unsaturated zone is often neglected and the water table is considered as an upper boundary of the saturated groundwater flow domain. Only if such a simplification is used, it becomes necessary to specify the water table as a boundary condition.

When the position of the water table is known, the water table condition is straightforward, because the pressure at the water table is zero and the following boundary condition applies:

$$h\left(x_w, y_w, z_w, t\right) = z_w, \tag{2.17}$$

where (x_w, y_w, z_w) represents known coordinates of a point on the water table. For instance, this approach was used by Tóth (1962, 1963), who assumed sloping and sinusoidally undulating water table configurations to describe regional groundwater flow systems.

However, in most practical cases, the position of the water table is not known beforehand but forms part of the solution. Hence, in such a case, Equation 2.17 becomes

$$h_w = h\left(x, y, z_w, t\right) = z_w\left(x, y, t\right), \tag{2.18}$$

and one more condition needs to be specified in order to solve the groundwater flow equation. This condition can be obtained by expressing the water balance resulting from the groundwater flows crossing the water table and the change in storage due to movement of the water table, as shown in Figure 2.1. In this figure, we depict the water table position at times t and $t+\Delta t$. It is important to note that the water table elevation $z_w(x, y, t)$ is only a function of the horizontal coordinates x and y and of the time t, but not of the vertical coordinate z. Hence, the vertical displacement of the water table is Δz_w, and, if $\mathbf{n} = (n_x, n_y, n_z)$ is the unit normal to the water table, the normal displacement of the water table becomes $n_z \Delta z_w$.

This displacement results in a change in groundwater storage, ΔV [L^3], given by

$$\Delta V = S_y n_z \Delta z_w \Delta A, \tag{2.19}$$

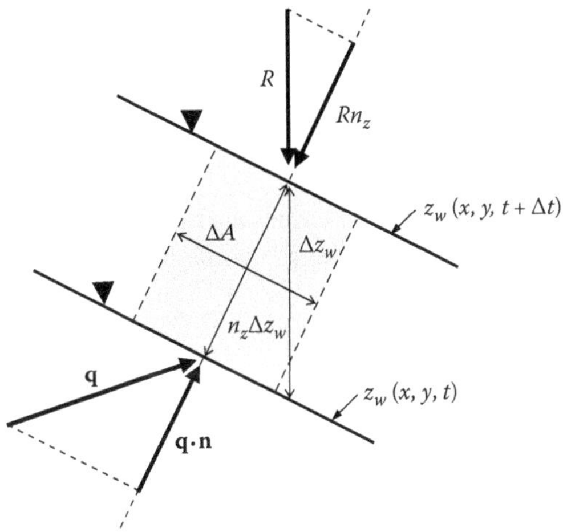

FIGURE 2.1 Boundary condition for a moving water table.

where ΔA [L^2] is an elementary area on the water table (Figure 2.1) and S_y [–] is the specific yield, which expresses the volume of water stored per volume of porous medium by a change in the water table position (see Section 3.4 for a discussion of specific yield). The change in storage, ΔV, should be balanced by the groundwater flows at the water table. Water that arrives at the water table from above via the unsaturated zone is denoted as recharge, R [L/T], and is usually assumed to be vertically downward (see Section 3.4), so that the normal component entering the water table is given by Rn_z, while in the saturated zone below the water table the normal component of the groundwater flow is given by $q_n = \mathbf{q}\cdot\mathbf{n}$ (Figure 2.1). Hence, mass balance implies that a change in storage equals inflow minus outflow, or

$$\Delta V = \left(\mathbf{q}\cdot\mathbf{n} + Rn_z\right)\Delta A \Delta t. \tag{2.20}$$

Combining Equations 2.19 and 2.20, simplifying and taking the limit for $\Delta t \to 0$, yields

$$S_y \frac{\partial z_w}{\partial t} n_z = \mathbf{q}\cdot\mathbf{n} + Rn_z. \tag{2.21}$$

An expression for the outer unit normal to the water table can be obtained by taking the gradient of $z - z_w$, i.e.,

$$\mathbf{n} = \frac{\nabla\left(z - z_w\right)}{\left|\nabla\left(z - z_w\right)\right|} = \frac{\left(-\partial z_w/\partial x, -\partial z_w/\partial y, 1\right)}{\left|\nabla\left(z - z_w\right)\right|}, \tag{2.22}$$

and when substituted into Equation 2.21 it results in

$$S_y \frac{\partial z_w}{\partial t} = -q_x \frac{\partial z_w}{\partial x} - q_y \frac{\partial z_w}{\partial y} + q_z + R. \tag{2.23}$$

This is a kinematic equation that describes the rate of change of the water table position. From Equation 2.18 and the chain rule for differentiation, it follows

$$\frac{\partial h_w}{\partial t} = \frac{\partial z_w}{\partial t}\left(1 - \frac{\partial h_w}{\partial z}\right), \quad \frac{\partial h_w}{\partial x} = \frac{\partial z_w}{\partial x}\left(1 - \frac{\partial h_w}{\partial z}\right), \quad \frac{\partial h_w}{\partial y} = \frac{\partial z_w}{\partial y}\left(1 - \frac{\partial h_w}{\partial z}\right), \tag{2.24}$$

where $\partial h_w/\partial t = \partial h/\partial t(x, y, z_w, t)$ and similarly for the other terms, so that in combination with Darcy's law, Equation 2.23 can be written as

$$S_y \frac{\partial h_w}{\partial t} = K\left[\left(\frac{\partial h_w}{\partial x}\right)^2 + \left(\frac{\partial h_w}{\partial y}\right)^2 + \left(\frac{\partial h_w}{\partial z}\right)^2\right] - (K + R)\frac{\partial h_w}{\partial z} + R, \tag{2.25}$$

which is the required additional boundary condition for the head at the water table. Unfortunately, this equation is nonlinear, which severely restricts applications in practice. Often this boundary condition is simplified by neglecting the nonlinear terms on the right-hand side, hence

$$S_y \frac{\partial h_w}{\partial t} = -(K + R)\frac{\partial h_w}{\partial z} + R. \tag{2.26}$$

This approximation is acceptable if the gradient of the head at the water table, $|\nabla h_w|$, is small. Equation 2.26 was used, for instance, by Neuman (1972) to derive an analytical solution for groundwater flow toward a well in a phreatic aquifer (without recharge).

There has been considerable discussion (and confusion) about the boundary condition given by Equation 2.25 and the related kinematic equation (Equation 2.23), e.g., Batu (2010), Bear and Cheng (2010), Jamali and Seymour (2011), Neuman (2011), Strack and Verruijt (2011), and Wang (2011).

To obtain numerical solutions of groundwater flow involving a water table, the position of the groundwater table is usually determined by iteration. Starting with an estimated water table elevation, z_w, the groundwater flow equation is solved using boundary condition (2.25), or more often the linearized form (2.26), to obtain the head, h, everywhere in the flow domain including the head, h_w, on the estimated water table. Then boundary condition (2.17) is used to determine a new water table elevation, $z_w = h_w$, and so on until the iterations have converged. Another more popular approach is based on determination of the new water table elevation as the position where the head, h, is equal to the elevation, z, i.e., by using Equation 2.18. However, in this approach, the first guess of the water table position has to be chosen above the expected position. Another possibility is to make use of the kinematic condition, given by Equation 2.23, to determine the water table elevation by iteration. Starting with a guessed water table elevation, z_w, boundary condition (2.17) is used to determine the head on the water table, $h_w = z_w$. With this head condition, the groundwater flow equation can be solved, which enables to determine the flux components on the phreatic surface. Substitution of these fluxes into the kinematic equation (2.23) makes it possible to improve the water table elevation and to iterate until convergence is achieved.

2.5 INVERSE PROBLEMS

In the previous sections, we have assumed that the hydraulic conductivity, specific storage, and specific yield are known beforehand. However, in most practical problems, the soil properties are not known very well, because these are difficult to ascertain accurately as appropriate measurement techniques can be tedious, costly, and time-consuming. Moreover, ground layers are heterogeneous and properties may exhibit considerable spatial variation. This is especially so for the hydraulic conductivity, which plays a dominant role in most groundwater flow problems. Therefore, one can try to infer the values and spatial distribution of the soil properties by so-called inverse modeling, i.e., by fitting the parameters to achieve a certain degree of correspondence between the model simulations and observations in the field. Field data are usually groundwater heads observed in piezometers or observation wells, while less attention is given to observations of groundwater fluxes, and model calibration is commonly achieved by minimizing the differences with model predictions using manual or automatic optimization techniques. However, there are many challenges and pitfalls in inverse modeling because parameter estimates may be unrealistic, nonunique, or subjected to large uncertainty (e.g., Sun, 2004; Hill and Tiedeman, 2007; Doherty and Hunt, 2010; Doherty et al., 2010).

In 1980, Calderón published a widely influential paper on the determination from boundary measurement of the spatially varying conductivity in a conducting medium. More precisely, he posed the following problem. Consider a modeling domain with spatially varying conductivity function $K(x,y,z)$ and two boundary conditions, both head and flux conditions (both Dirichlet and

Neumann conditions). Decide whether the conductivity function is uniquely determined by these head and flux boundary conditions. And, assuming this is the case, reconstruct function $K(x, y, z)$ from these boundary conditions. Calderón (1980) gave also arguments that a unique solution exists if function $K(x, y, z)$ honors some smoothness conditions (e.g., differentiability). Originally, Calderón (1980) considered problems governed by the electrical potential equation, which has numerous applications ranging from biomedical imaging to nondestructive testing and geophysics (Borcea, 2002). No general formal proof of this conjecture has been found, but application to a number of practical problems shows it to be trustworthy. Using head and flux instead of electrical potential and current, a similar approach can be applied to tackle parameter reconstruction in groundwater flow modeling.

In inverse groundwater modeling with Calderón's technique, one intends to determine the hydraulic conductivity from head and flux observations. When all head and normal flux components on the boundary of a groundwater domain are given for an incompressible (not necessarily steady) flow problem, the hydraulic conductivity in each point of the flow domain can be obtained by iteration. Starting with a guessed distribution, $K(x, y, z)$, groundwater flow equation (2.6) is solved using only the given flux boundary conditions to obtain a first solution, h_1. From this solution, groundwater fluxes, $\mathbf{q}_1$, can be derived in each point of the flow domain. Next, groundwater flow equation (2.6) is solved again using only the given head boundary conditions to obtain a second solution, h_2, which enables to determine head gradients, ∇h_2, in each point of the flow domain. Then the conductivity function is improved using Darcy's law

$$K = \frac{|\mathbf{q}_1|}{|\nabla h_2|}, \tag{2.27}$$

and the procedure is started anew until convergence of the "Darcy residual," $\iiint (K^{-1/2}\mathbf{q}_1 + K^{1/2}\nabla h_2)^2 \mathrm{d}x\mathrm{d}y\mathrm{d}z$, to a (local) minimum is reached (Kohn and Vogelius, 1987). However, in practice, it will be nearly impossible to observe all heads and normal flux components on the boundary of the flow domain. Nevertheless, both head and flux may be known on some parts of the boundary. For instance, on the phreatic surface two conditions may be known, at least approximately: the head, $h_w = z_w$, and the recharge flux, R. Furthermore, for a piezometer (a monitoring well), two internal boundary conditions are known: the observed head, h_b, and, in addition, a zero flux, $q_b = 0$; this is useful information, which is often ignored or unconsciously taken into account. Other examples are observed pumping rates and heads in wells, which may be considered as internal boundaries, water levels, and drainage rates in streams or drains, and so on. All such information can form the basis for a partial calibration. Comparison of the results obtained from a head-specified model with those of a flux-specified model may give insight into the range and spatial distribution of the hydraulic conductivity. The differences in heads and fluxes obtained from the two models can be used to guide the calibration, i.e., to modify hydraulic conductivity in such a way that the models honor both the known head and the flux conditions, e.g., El-Rawy et al. (2015).

3 Hydraulic Approach to Groundwater Flow

3.1 CONCEPT

In general, groundwater flow is three-dimensional, but due to the geometry of ground layers and differences in hydraulic conductivities, actual groundwater flows tend to be concentrated in certain directions, as illustrated in Figure 3.1. The reason for this is that ground layers usually extend horizontally over large distances, while the vertical dimensions are rather restricted. Furthermore, due to the large variety in conductive properties, ground layers can be grouped into three classes: pervious formations or aquifers, semipervious formations or aquitards, and impervious formations (Figure 3.1a). Because water flows through a porous medium along the path of least resistance, groundwater moves predominantly longitudinally in pervious formations and transversely in semipervious layers. Hence, due to the horizontal layering, groundwater flow is essentially horizontal in aquifers and vertical in aquitards, as depicted schematically in Figure 3.1b. Therefore, instead of using the three-dimensional equations as presented in Section 3.2, a simplified mathematical description of groundwater flow is possible, when horizontal flow components are considered to be dominant in aquifers and vertical components in aquitards, as shown in Figure 3.1c. This concept is termed the hydraulic approach to groundwater

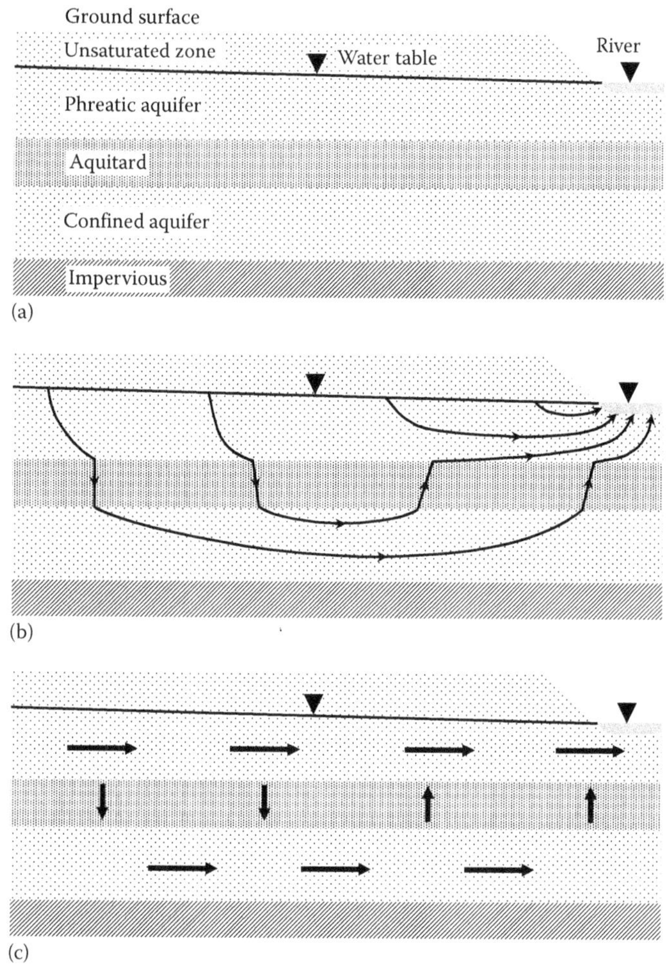

FIGURE 3.1 Occurrence and movement of groundwater: (a) types of hydrogeological layers, (b) groundwater flow paths, and (c) predominant groundwater flow directions.

flow, because the mathematical equations more or less resemble the equations describing water flow in pipes or channels (piston flow); see Polubarinova-Kochina (1962) and Bear (1972, 1979). In case of confined aquifers, this approach is also called the Dupuit approximation, and in the case of a phreatic aquifer it is called the Dupuit–Forchheimer approximation, after the two scientists who pioneered this approach in the second half of the nineteenth century.

3.2 EQUATIONS OF MOTION

In order to have a clear understanding of the basic concepts of the hydraulic approach, it is convenient to start with the equation of motion, before discussing the continuity equation. Consider a confined aquifer, bounded by a bottom surface, $b(x,y)$, and an upper surface, $s(x,y)$, such that its thickness is given by $D=s-b$, as shown in Figure 3.2a. Due to the geometry, vertical groundwater flow is very small compared to horizontal flow, and vertical Darcy's law, $\partial h/\partial z=-q_z/K_v$ (Equation 1.29), may be replaced by the Dupuit approximation, $\partial h/\partial z=0$, implying that the head, h, is independent of z. Under this assumption, the total horizontal volumetric flow rate of the groundwater in the x direction over the entire thickness of the aquifer can be calculated as

$$Q_x(x,y,t)=\int_b^s q_x(x,y,z,t)\mathrm{d}z=-\int_b^s K_h(x,y,z)\mathrm{d}z\frac{\partial h(x,y,t)}{\partial x}=-T(x,y)\frac{\partial h(x,y,t)}{\partial x},\tag{3.1}$$

where Q_x is the x component of the hydraulic groundwater flux, $\mathbf{Q}=(Q_x,Q_y)$, with dimensions volume per time per length [L^3/TL = L^2/T], and $T(x,y)=\int_b^s K_h(x,y,z)\mathrm{d}z$ is the transmissivity [L^2/T], which expresses the overall horizontal conductance of a confined aquifer. If the horizontal hydraulic conductivity, K_h, does not depend on z, $T=K_hD$. Similarly, the y component of the hydraulic flux is given by

$$Q_y(x,y,t)=-T(x,y)\frac{\partial h(x,y,t)}{\partial y},\tag{3.2}$$

Equations 3.1 and 3.2 describe overall (depth-averaged) horizontal flow of groundwater in a confined aquifer, and in the hydraulic approach these equations replace horizontal Darcy's laws, $q_x=-K_h\partial h/\partial x$ and $q_y=-K_h\partial h/\partial y$ (Equation 1.32).

The case of a phreatic aquifer is shown in Figure 3.2b. The aquifer is bounded from below by its base, $b(x,y)$, and from above by the soil surface, $s(x,y)$. However, when the unsaturated zone is ignored, the upper boundary of the phreatic aquifer becomes the water table position, $z_w(x,y,t)$. Similar considerations apply as for a confined aquifer, so vertical groundwater flow is negligible compared to horizontal flow and the Dupuit approximation applies

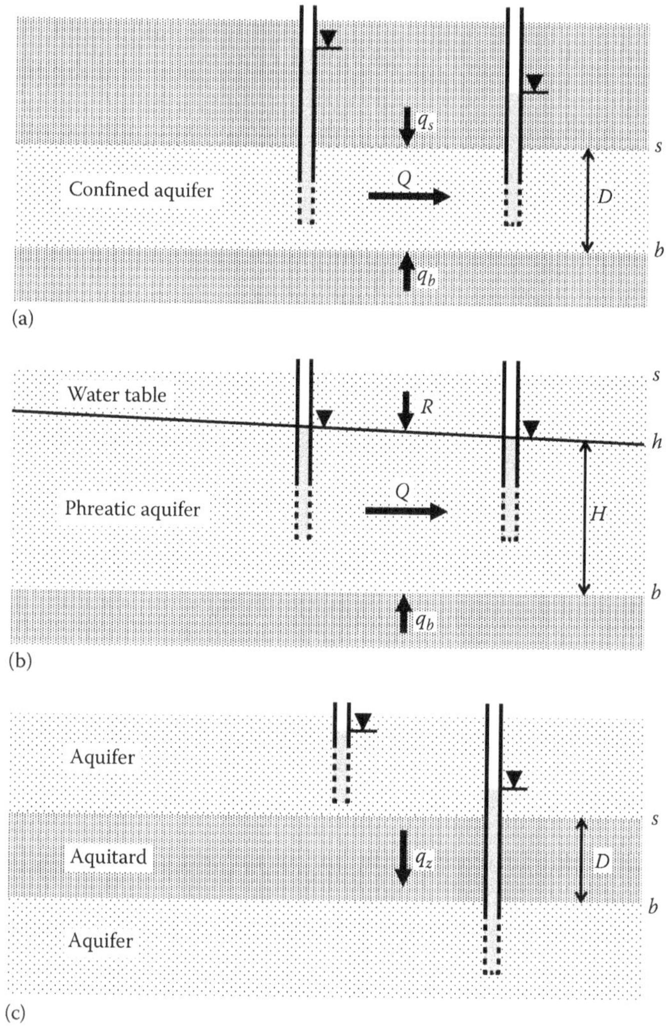

FIGURE 3.2 Hydraulic approach to groundwater flow: (a) horizontal flow in a confined aquifer, (b) horizontal flow in a phreatic aquifer, and (c) vertical flow in an aquitard.

in which vertical Darcy's law is replaced with $\partial h/\partial z = 0$. Hence, the head, h, is independent of z and, according to the boundary condition given by Equation 2.18, coincides with the position of the water table, i.e., $h = z_w$. Moreover, because in Equation 2.24 the term $(1 - \partial h_w/\partial z)$ vanishes, also $\partial h/\partial x = \partial z_w/\partial x$, $\partial h/\partial y = \partial z_w/\partial y$, and $\partial h/\partial t = \partial z_w/\partial t$ coincide on the water table.

Consequently, the total flow of groundwater passing horizontally through the aquifer in the x direction is given by

$$Q_x(x,y,t) = \int_b^h q_x(x,y,z,t)\mathrm{d}z = -\int_b^h K_h(x,y,z)\mathrm{d}z\frac{\partial h(x,y,t)}{\partial x} = -K_e(x,y)(h-b)\frac{\partial h}{\partial x}, \tag{3.3}$$

where $K_e = \int_b^h K_h \mathrm{d}z/(h-b)$ is the effective or mean horizontal hydraulic conductivity and is equal to K_h, if K_h does not depend on z. However, ground layers are generally heterogeneous, so that K_e needs to be determined by averaging; this is called upscaling or homogenization and requires careful consideration of the hydrogeological and stratigraphic characteristics of the subsurface and of the averaging procedure. Theoretical and mathematical aspects are discussed by, e.g., Zijl and Nawalany (1993) and Rubin (2003). Also, transmissivity can be defined similarly to a confined aquifer, but this makes less sense, because the thickness of the phreatic aquifer changes with changes in elevation of the water table, so that the transmissivity is not constant but changes in time.

An expression similar to Equation 3.3 can be obtained for the flow component in the y direction as follows:

$$Q_y(x,y,t) = -K_e(x,y)(h-b)\frac{\partial h}{\partial y}. \tag{3.4}$$

Equations 3.3 and 3.4 are due to Dupuit–Forchheimer, who postulated that groundwater flow in a phreatic aquifer is proportional to the slope of the groundwater table, and replace horizontal Darcy's laws, but unfortunately are nonlinear. A simplification is possible for a phreatic aquifer with a horizontal base, using the elevation of the water table above the base, $H(x,y,t) = z_w(x,y,t) - b = h(x,y,t) - b$, which, thanks to the Dupuit approximation, is also the head measured from the base of the aquifer. Therefore, H may be used as the dependent state variable, and the hydraulic flux components can then be written as

$$Q_x = -K_e H\frac{\partial H}{\partial x} = -K_e\frac{\partial H^2/2}{\partial x}, \quad Q_y = -K_e H\frac{\partial H}{\partial y} = -K_e\frac{\partial H^2/2}{\partial h}. \tag{3.5}$$

Dupuit–Forchheimer's approximation is mostly applied in this form because the resulting equations are linear in $H^2/2$.

In case of an aquitard bounded by two aquifers, horizontal groundwater flow is very small compared to vertical flow and the horizontal flux components, q_x and q_y, may be ignored, as depicted in Figure 3.2c. If, in addition, the flow is assumed to be incompressible, mass balance (Equation 2.9) implies that $\partial q_z/\partial z=0$. Hence, the vertical flow passing through the aquitard, q_z, is independent of z and can be derived as follows:

$$h(x,y,z_s,t)-h(x,y,z_b,t)=\int_b^s \frac{\partial h}{\partial z}\mathrm{d}z=-\int_b^s \frac{q_z(x,y,t)}{K_v(x,y,z)}\mathrm{d}z=-\frac{q_z(x,y,t)}{C(x,y)}, \tag{3.6}$$

or

$$q_z(x,y,t)=-C(x,y)\left[h(x,y,z_s,t)-h(x,y,z_b,t)\right], \tag{3.7}$$

where $C=\left(\int_b^s K_v^{-1}\mathrm{d}z\right)^{-1}$ is the conductance, commonly called the leakage coefficient, with dimensions [T^{-1}], which expresses the overall leaking capacity of an aquitard allowing vertical groundwater transfer between two aquifers bounding the aquitard. If the vertical hydraulic conductivity, K_v, does not depend on z, $C=K_v/D$. Notice that the leakage coefficient depends on the harmonic average of the vertical hydraulic conductivity over the aquitard thickness, while the transmissivity of a confined or a phreatic aquifer depends on the arithmetic average of the horizontal hydraulic conductivity over the aquifer thickness.

3.3 FLOW EQUATION FOR A CONFINED AQUIFER

To obtain a flow equation for a confined aquifer based on the hydraulic approach, the three-dimensional flow equation is integrated over the total depth of the aquifer. For this, we write Equation 2.5 in the form of a mass balance equation as follows

$$S_s\partial h/\partial t+\partial q_x/\partial x+\partial q_y/\partial y+\partial q_z/\partial z=0, \tag{3.8}$$

and integrate along the vertical from the bottom plane, $b(x,y)$, to the upper surface, $s(x,y)$, i.e.

$$\int_b^s S_s\partial h/\partial t\mathrm{d}z+\int_b^s\left(\partial q_x/\partial x+\partial q_y/\partial y+\partial q_z/\partial z\right)\mathrm{d}z=0. \tag{3.9}$$

This can be worked out using Leibniz's rule for differentiation and assuming that the head, $h(x,y,t)$, does not depend on z, as

$$\int_b^s S_s \mathrm{d}z \frac{\partial h}{\partial t} + \frac{\partial}{\partial x}\int_b^s q_x \mathrm{d}z - q_x\Big|_s \frac{\partial s}{\partial x} + q_x\Big|_b \frac{\partial b}{\partial x} + \frac{\partial}{\partial y}\int_b^s q_y \mathrm{d}z - q_y\Big|_s \frac{\partial s}{\partial y} + q_y\Big|_b \frac{\partial b}{\partial y}$$
$$+ q_z\Big|_s - q_z\Big|_b = 0. \qquad (3.10)$$

Next, we note that $q_b = q_x|_b \partial b/\partial x + q_y|_b \partial b/\partial y - q_z|_b$ is the vertical flux through the underlying aquitard entering the aquifer from below, while $q_s = -q_x|_s \partial s/\partial x - q_y|_s \partial s/\partial y + q_z|_s$ is the vertical flux through the upper aquitard entering the aquifer from above, as shown in Figure 3.2b, with sign convention that inflow is considered positive irrespective of the direction along the z-axis (if no aquitards are present, q_b and q_s are zero), so that Equation 3.10 becomes

$$S\frac{\partial h}{\partial t} = -\frac{\partial Q_x}{\partial x} - \frac{\partial Q_y}{\partial y} + q_s + q_b, \qquad (3.11)$$

or

$$S\frac{\partial h}{\partial t} = \frac{\partial}{\partial x}\left(T\frac{\partial h}{\partial x}\right) + \frac{\partial}{\partial y}\left(T\frac{\partial h}{\partial y}\right) + q_s + q_b, \qquad (3.12)$$

where we have made use of the hydraulic motion Equations 3.1 and 3.2, and $S = \int_b^s S_s \mathrm{d}z$ is the aquifer storage coefficient or storativity, which is dimensionless and equal to $S_s D$ if the specific storage coefficient, S_s, does not depend on z. The storativity is a measure of the overall aquifer storage properties and can be defined as the volume of groundwater released by a confined aquifer per unit horizontal aquifer surface and per unit decline of hydraulic head, all in accordance with the hydraulic approach where variables are considered averaged or integrated over the vertical dimensions of the aquifer.

Compared to three-dimensional groundwater flow, described by Equation 2.5, it can be noted that Equation 3.12 contains no z dimension, which is the main simplification resulting from the hydraulic approach. Also, as a consequence, the needed porous medium properties are transmissivity and storativity, which are aquifer properties integrated over the vertical dimension. Therefore, Equation 3.12 is simpler and needs less detailed knowledge about the medium properties. Care should be taken when applying this equation in practice, because boundary conditions are required in accordance with the hydraulic approach. This implies that boundary conditions should be independent of the elevation, i.e., they should apply to the total aquifer thickness.

For instance, pumping by a fully penetrating well can be considered with the hydraulic approach, but not a partially penetrating well because this induces appreciable vertical head gradients in the aquifer.

3.4 FLOW EQUATION FOR A PHREATIC AQUIFER

A flow equation for a phreatic aquifer based on the hydraulic approach can be obtained similar as for a confined aquifer, but some complications arise because the upper boundary of a phreatic aquifer is a water table and changes in water table elevation have to be taken into account. The flow equation is obtained by integrating the three-dimensional flow Equation 3.8 over the depth of the aquifer, which now extends from the base, $b(x, y)$, to the water table, $z_w(x, y, t)$, as shown in Figure 3.2b, i.e.

$$\int_b^{z_w} S_s \partial h/\partial t \mathrm{d}z + \int_b^{z_w} \left(\partial q_x/\partial x + \partial q_y/\partial y + \partial q_z/\partial z\right)\mathrm{d}z = 0. \tag{3.13}$$

This can be worked out and assuming that the head, $h(x, y, t)$, does not depend on z, as

$$\int_b^{z_w} S_s \mathrm{d}z \frac{\partial h}{\partial t} + \frac{\partial}{\partial x}\int_b^{z_w} q_x \mathrm{d}z - q_x\big|_{z_w}\frac{\partial z_w}{\partial x} + q_x\big|_b\frac{\partial b}{\partial x} + \frac{\partial}{\partial y}\int_b^{z_w} q_y \mathrm{d}z - q_y\big|_{z_w}\frac{\partial z_w}{\partial y} + q_y\big|_b\frac{\partial b}{\partial y}$$
$$+ q_z\big|_{z_w} - q_z\big|_b = 0. \tag{3.14}$$

Next, we introduce the vertical flux through the underlying aquitard entering the aquifer from below, $q_b = q_x|_b \partial b/\partial x + q_y|_b \partial b/\partial y - q_z|_b$, and make use of the kinematic Equation 2.23 for the water table, to obtain

$$\int_b^{z_w} S_s \mathrm{d}z \frac{\partial h}{\partial t} = -\frac{\partial Q_x}{\partial x} - \frac{\partial Q_y}{\partial y} + R - S_y\frac{\partial z_w}{\partial t} + q_b. \tag{3.15}$$

Finally, we note that due to the Dupuit assumption, $z_w = h_w = h$, this becomes

$$\left[S_y + S_s(h-b)\right]\frac{\partial h}{\partial t} = -\frac{\partial Q_x}{\partial x} - \frac{\partial Q_y}{\partial y} + R + q_b, \tag{3.16}$$

or

$$\left[S_y + S_s\left(h-b\right)\right]\frac{\partial h}{\partial t} = \frac{\partial}{\partial x}\left[K_e\left(h-b\right)\frac{\partial h}{\partial x}\right] + \frac{\partial}{\partial y}\left[K_e\left(h-b\right)\frac{\partial h}{\partial y}\right] + R + q_b, \tag{3.17}$$

where we have made use of the hydraulic motion Equations 3.3 and 3.4 and assumed that S_s does not depend on z. Also in this equation there are no terms with respect to the z dimension, which is again the main advantage of the hydraulic approach. However, the equation is nonlinear, which complicates practical applications and finding exact analytical solutions. Often the storage term is simplified by ignoring the effect of aquifer and water compressibility, $S_s(h-b)$, because this is usually much smaller than the specific yield S_y. Also, care should be taken with respect to boundary conditions, because these have to respect the hydraulic approach and consequently be independent of elevation. Consider for instance the very common boundary condition of a river in contact with a phreatic aquifer. The water level in the river can only be used as a head boundary condition, when the river is wide and deep such that the groundwater underneath the river is effectively having the same head. If the surface water body is narrow and shallow, like drains or ditches, there can be significant vertical flow components resulting in a head difference between the surface water and the groundwater underneath. In practice, such a situation can be treated using a third-type boundary condition (or Robin boundary condition), where a boundary resistance is introduced that takes into account the effect of converging groundwater flow toward the drain or ditch.

A simplification is possible for phreatic aquifers with a horizontal base, using the elevation of the water table above the base, $H(x,y,t)=z_w(x,y,t)-b$. Thanks to the Dupuit approximation, H is also the head. Therefore, H may be used as the dependent variable so that Equation 3.17 can be written as

$$S_y\frac{\partial H}{\partial t} = \frac{\partial}{\partial x}\left[K_e\frac{\partial H^2/2}{\partial x}\right] + \frac{\partial}{\partial y}\left[K_e\frac{\partial H^2/2}{\partial y}\right] + R + q_b, \tag{3.18}$$

where also aquifer and water compressibility effects have been ignored. This is called the Boussinesq equation. The equation is still nonlinear, but in case of steady-state conditions becomes

$$\frac{\partial}{\partial x}\left[K_e\frac{\partial H^2/2}{\partial x}\right] + \frac{\partial}{\partial y}\left[K_e\frac{\partial H^2/2}{\partial y}\right] + R + q_b = 0, \tag{3.19}$$

which is linear in $H^2/2$. Several useful solutions for groundwater flow in a phreatic aquifer have been derived with this equation (e.g., Polubarinova-Kochina, 1962; Bear, 1972, 1979).

We end this section with some remarks about the recharge, R, and the specific yield, S_y, because these are the result of flow processes taking place in the unsaturated zone above the water table. These processes are rather complex and, hence, have to be taken into account in a simplified way, unless unsaturated flow is taken into consideration, which is exactly what one would like to avoid by applying the hydraulic approach. Recharge depends on water transfers such as infiltration and evapotranspiration taking place at the soil surface and in the upper part of the unsaturated zone and the subsequent redistribution and drainage of the remaining water toward the water table. All of these processes are difficult to describe in detail, but because groundwater flow is a slow process it is reasonable to consider an average flux for the recharge derived from mass balance considerations in the unsaturated zone over long periods of time. Also, the exact evaluation of specific yield is not that simple. The situation is shown schematically in Figure 3.3. If the water table drops by an amount Δz_w, the change in storage depends upon the moisture distribution above the water table before and after the drop in water table elevation, and also time will have an influence because the released water in the unsaturated

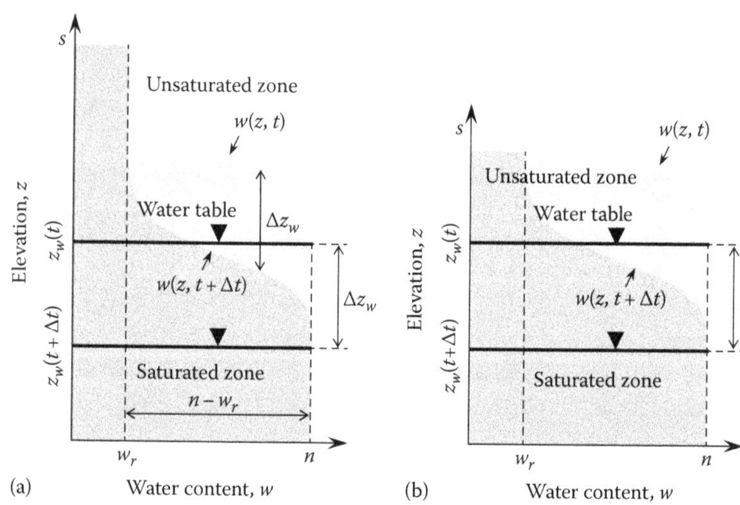

FIGURE 3.3 Storage and release of groundwater in a phreatic aquifer due to changes in water table position: (a) deep water table and (b) shallow water table.

zone has to percolate downward to the water table. Hence, the change in storage needs to be approximated in a more simple way. This can be achieved as follows. Consider the situation depicted in Figure 3.3a, where the soil surface is high above the water table. Under normal average conditions, the distribution of the water content, w, will vary with elevation above the water table. It will be equal to the porosity at the water table and generally will decrease with height above the water table, gradually approaching a residual water content, w_r, which is characteristic for the unsaturated zone far above the water table. If the water table elevation is changed by an amount Δz_w, and given sufficient time, the moisture distribution will eventually become similar as before, only shifted over a distance Δz_w. Hence, the volumetric difference in storage can be calculated as

$$\int_{z_w}^{s} \frac{\partial w}{\partial t} dz = \int_{z_w}^{s} \lim_{\Delta t \to 0} \left[\frac{w(z,t+\Delta t) - w(z,t)}{\Delta t} \right] dz = \lim_{\Delta t \to 0} \left[(n - w_r) \frac{\Delta z_w}{\Delta t} \right]$$
$$= (n - w_r) \frac{\partial z_w}{\partial t} = S_y \frac{\partial z_w}{\partial t}, \qquad (3.20)$$

where $S_y = n - w_r$ [—] is defined as the specific yield, i.e., the amount of water released per volume of porous medium when changing from the saturated state to the unsaturated state high above the water table. In reality, things are not that simple. First, it could be possible that the soil surface is not situated far above the water table, as shown in Figure 3.3b, such that the mathematical derivation given above is not valid. Second, retardation effects will occur, because conductivities in the unsaturated zone are generally much smaller than in the saturated zone, and, consequently, water movement in the unsaturated zone lags behind saturated groundwater flow. Third, soil heterogeneity complicates the concept of residual water content and equilibrium moisture distributions in the unsaturated zone. Hence, in practice, the change in storage will usually be smaller. Therefore, Equation 3.20 remains only useful when the concept of specific yield is extended to match any particular situation. This gross simplification is justified in view of the fact that with the hydraulic approach—and also in the three-dimensional approach with the phreatic surface conditions (Section 2.4)—groundwater flow is described in a simplified way. However, the concept of specific yield has a clear physical interpretation, similar as storativity, i.e., the volume of water released by a phreatic aquifer per unit horizontal aquifer surface and per unit decline of the water table. Of course, the position of the water table, the speed of water table decline, and the soil heterogeneity play an important role in this.

3.5 FLOW EQUATION FOR AN AQUITARD

For obtaining a flow equation for an aquitard, it is sufficient to note that the flow is considered to be strictly vertical, so that the flow equation can be derived from the three-dimensional groundwater flow equation, Equation 2.5, by only considering the z component of the flow as follows:

$$S_s \frac{\partial h}{\partial t} = \frac{\partial}{\partial z}\left(K_v \frac{\partial h}{\partial z} \right). \tag{3.21}$$

This equation describes flow through an aquitard, while the heads of the aquifers bounding the aquitard can be used as boundary conditions. However, in practice, this equation is not used very much. Usually, an additional simplifying assumption is made, namely that the flow is incompressible, so that

$$\frac{\partial}{\partial z}\left(K_v \frac{\partial h}{\partial z} \right) = 0. \tag{3.22}$$

This means that the flow is in quasi steady state and the solution is given by Equation 3.7, i.e., the flow adjusts immediately to changes in the aquifer heads above and below the aquitard. This simplification makes vertical flows through aquitards easy to evaluate, so that these can be incorporated as inputs or outputs in above or underlying aquifers. However, the approach is not very accurate, because conductivities of aquitards are much smaller than for aquifers, so that, in general, the flow through an aquitard lags behind the flow in the bounding aquifers and time and storage effects can be important (Frind, 1983). However, lack of knowledge of aquitard properties often forces us to use a simplified approach.

3.6 EXTENSION OF THE HYDRAULIC APPROACH

Although the hydraulic approach is two-dimensional in the horizontal coordinates x and y, the flow in the aquifer can still to be "pictured" as three-dimensional by deriving an approximate expression for the vertical flux component, q_z. At first glance this seems to contradict the Dupuit approximation, because

$\partial h/\partial z=0$ makes it impossible to determine the vertical flux by the vertical Darcy's law (Equation 1.32). However, instead of using the vertical Darcy's law, q_z can be derived by closure of the three-dimensional groundwater flow Equation 2.5 (Strack, 1984):

$$\frac{\partial q_z}{\partial z}=\frac{\partial}{\partial x}\left(K_h\frac{\partial h_0}{\partial x}\right)+\frac{\partial}{\partial y}\left(K_h\frac{\partial h_0}{\partial y}\right)-S_s\frac{\partial h_0}{\partial t}, \tag{3.23}$$

where $h_0(x,y,t)$ is the solution derived with the hydraulic approach. By integrating and assuming incompressible flow, q_z is obtained as

$$q_z=q_z\big|_b+\int_b^z\left[\frac{\partial}{\partial x}\left(K_h\frac{\partial h_0}{\partial x}\right)+\frac{\partial}{\partial y}\left(K_h\frac{\partial h_0}{\partial y}\right)\right]\mathrm{d}z, \tag{3.24}$$

which can be used to improve the groundwater head as

$$h(x,y,z,t)=h_0(x,y,t)-\int\frac{q_z}{K_v}\mathrm{d}z+C, \tag{3.25}$$

where C is an integration constant to be set by the boundary conditions.

For instance, for a confined aquifer, $h_0(x,y,t)$ is determined with the Dupuit Equation 3.12 and inserted into Equation 3.24 to obtain q_z, which, if K_h does not depend upon z, $b(x,y)$ and $s(x,y)$ vary only mildly, and assuming incompressible flow, reduces to

$$q_z=q_b-(q_s+q_b)(z-b)/D, \tag{3.26}$$

which simply states that the vertical flux q_z varies linearly between q_b and $-q_s$. The improved groundwater head becomes

$$h=h_0-q_bz/K_v+(q_s+q_b)\left(\tfrac{1}{2}z^2-bz\right)/K_vD+C, \tag{3.27}$$

where we have assumed that K_v does not depend upon z. In case of a phreatic aquifer, the procedure is similar, although the water table comes into play. The first estimate of the groundwater head, $h_0(x,y,t)$, is now obtained with Equation 3.17 and when inserted in Equation 3.24, assuming K_h and K_v do not depend upon z and $b(x,y)$ varies only mildly, results in

$$q_z=q_b-\left[R-S_y\partial h_0/\partial t+q_b+K_h(\partial h_0/\partial x)^2+K_h(\partial h_0/\partial y)^2\right](z-b)/(h_0-b), \tag{3.28}$$

and

$$h = h_0 - q_b z / K_v + \left[R - S_y \partial h_0 / \partial t + q_b + K_h \left(\partial h_0 / \partial x \right)^2 + K_h \left(\partial h_0 / \partial y \right)^2 \right] \times \left(\tfrac{1}{2} z^2 - bz \right) / K_v D + C, \tag{3.29}$$

where the term $-S_y \partial h_0 / \partial t$ can be omitted for steady-state conditions.

This procedure can be extended as a perturbation method, i.e., a hierarchy of iterative analytical approximations from which an asymptotic solution can be found for any three-dimensional groundwater flow problem (Bear, 1988; Zijl and Nawalany, 1993), at least in theory. In practice, this is rather unfeasible, because the iterations become increasingly complicated and the methodology does not explain how boundary conditions that are strictly three-dimensional have to be included, as a partially penetrating well or a seepage face.

Examples of Solutions

4

4.1 DUPUIT DISCHARGE FORMULA

Consider steady flow through a pervious homogeneous and isotropic dam with vertical walls and impervious horizontal base, as shown in Figure 4.1. The dam separates two reservoirs, so that the flow will be directed from the reservoir with the highest water level on the left to the reservoir with the lower level on the right and bounded vertically by the impervious base and the water table. In addition, there will be a seepage face at the right-hand side above the reservoir where groundwater seeps out freely in the atmosphere under zero pressure.

The flow equation, given by Equation 2.8, reduces to the two-dimensional Laplace equation

$$\frac{\partial^2 h}{\partial x^2} + \frac{\partial^2 h}{\partial z^2} = 0. \tag{4.1}$$

The boundary conditions are as follows. For the left boundary

$$h(0,z) = h_1, \tag{4.2}$$

where h_1 is the water depth in the left reservoir. For the right boundary, at the contact between dam and reservoir

$$h(L,z \le h_2) = h_2, \tag{4.3}$$

where h_2 is the water depth in the right reservoir and L the width of the dam, while for the seepage face above the reservoir

$$h(L,z > h_2) = z. \tag{4.4}$$

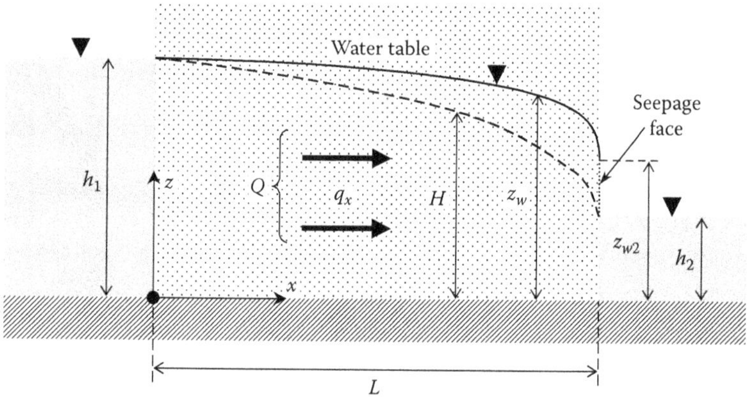

FIGURE 4.1 Steady-state flow through a porous dam with vertical walls, situated between two reservoirs with constant water levels; the flow is vertically contained between the impervious base and the water table and partially exits in the atmosphere through a seepage face; the dashed line represents the approximate position of the water table as obtained from the Dupuit equation.

For the impervious base

$$\left.\frac{\partial h}{\partial z}\right|_{z=0} = 0. \tag{4.5}$$

For the water table Equation 2.25 applies, which in this case becomes

$$\left(\frac{\partial h_w}{\partial x}\right)^2 + \left(\frac{\partial h_w}{\partial z}\right)^2 - \frac{\partial h_w}{\partial z} = 0, \tag{4.6}$$

where $h_w = h(x, y, z_w, t) = z_w(x, y, t)$ and recharge is ignored. An exact solution obtained by conformal mapping is given by Polubarinova-Kochina (1962), but it is very complicated and hard to apply in practice. Therefore, we present here the hydraulic approximation, so that Equation 3.19 applies, which in this case reduces to

$$\frac{\partial^2 H^2}{\partial x^2} = 0, \tag{4.7}$$

where $H = h$ because the base of the dam coincides with the zero elevation plane. The solution of Equation 4.7 is

$$H^2 = C_1 x + C_2, \tag{4.8}$$

where C_1 and C_2 are integration constants that have to be determined from the boundary conditions. Because the water table starts at h_1, the water level of the left reservoir at $x=0$, it follows that $C_2 = h_1^2$. At the right-hand side, there is a problem, because we do not know the position of the water table due to the presence of the seepage face. Hence, we are forced to ignore the seepage face, and impose $H=h_2$ for $x=L$, which enables to determine C_1, resulting in the following solution:

$$H^2 = \left(h_2^2 - h_1^2\right)\left(\frac{x}{L}\right) + h_1^2. \tag{4.9}$$

This equation implies that the water table has a parabolic shape, as depicted by the dashed line in Figure 4.1, which obviously deviates from reality (although differences will be small if L is large compared to h_1). The amount of flow, Q, from the left reservoir to the right reservoir can be calculated with Equation 3.5, yielding

$$Q = -K\frac{\partial H^2/2}{\partial x}. \tag{4.10}$$

Using Equation 4.9, this results in

$$Q = K\frac{h_1^2 - h_2^2}{2L}. \tag{4.11}$$

This is the Dupuit equation, which is extensively used in practice to calculate groundwater flow in phreatic aquifers.

One would be tempted to question its validity or accuracy in view of the Dupuit approximation that the flow is assumed to be horizontal, which obviously is physically impossible. To investigate this question, we present another derivation that is not based on the Dupuit assumption. To do so, we return to the two-dimensional flow problem, given by Equations 4.1 through 4.6, and calculate the flow from integrating the horizontal flux component, q_x, along a vertical from the base of the dam to the water table elevation, z_w, hence

$$Q = \int_0^{z_w} q_x \mathrm{d}z = -K\int_0^{z_w}\left(\frac{\partial h}{\partial x}\right)\mathrm{d}z. \tag{4.12}$$

When the integration and derivative are interchanged, this becomes

$$Q = -K\frac{\partial}{\partial x}\left(\int_0^{z_w} h\mathrm{d}z - \tfrac{1}{2}z_w^2\right). \tag{4.13}$$

Since, according to the water balance, Q is constant, integration with respect to x gives

$$Qx = -K\left(\int_0^{z_w} h\mathrm{d}z - \tfrac{1}{2}z_w^2\right) + C_3. \tag{4.14}$$

The integration constant, C_3, is determined by equating terms at $x=0$, for which we make use of the boundary condition given by Equation 4.2, which also implies $z_w(x=0)=h_1$, so that

$$C_3 = K\left(\int_0^{h_1} h_1\mathrm{d}z - \tfrac{1}{2}h_1^2\right) = K\left(h_1^2 - \tfrac{1}{2}h_1^2\right) = \tfrac{1}{2}Kh_1^2. \tag{4.15}$$

Equation 4.14 now becomes

$$Qx = \tfrac{1}{2}Kh_1^2 - K\left(\int_0^{z_w} h\mathrm{d}z - \tfrac{1}{2}z_w^2\right). \tag{4.16}$$

Next, we use the boundary conditions at $x=L$, given by Equations 4.3 and 4.4, yielding

$$\begin{aligned} QL &= \tfrac{1}{2}Kh_1^2 - K\left(\int_0^{h_2} h_2\mathrm{d}z + \int_{h_2}^{z_{w2}} z\mathrm{d}z - \tfrac{1}{2}z_{w2}^2\right) = \tfrac{1}{2}Kh_1^2 - K\left(h_2^2 + \tfrac{1}{2}z_{w2}^2 - \tfrac{1}{2}h_2^2 - \tfrac{1}{2}z_{w2}^2\right) \\ &= \tfrac{1}{2}K\left(h_1^2 - h_2^2\right), \end{aligned} \tag{4.17}$$

where $z_{w2}=z_w(L)$ is the water table position and also the top position of the seepage face at the right boundary. Surprisingly, the final result, Equation 4.17, yields again the Dupuit equation (4.11), which does not depend upon z_{w2} although the seepage face contributes to the total flow. This remarkable finding is due to Charny (1951). Hence, it appears that although the shape of the water table is not correctly predicted by the Dupuit approach, the equation for the flux is nevertheless exact. Obviously, the errors introduced by the parabolic shape of the water table and the disregard of the seepage face cancel out.

4.2 HVORSLEV SLUG TEST EQUATION

Consider a piezometer, with a screen of length L and radius R, installed in an aquifer, as depicted in Figure 4.2a. We assume that the screen is small compared to the aquifer dimensions, so that seen from the piezometer screen the aquifer can be considered infinite in all directions. The groundwater head h_0 is revealed by the water level in the piezometer. We will assume that this head is present everywhere in the aquifer. The question arises what will be the groundwater inflow through the screen into the piezometer if the water level inside the piezometer is artificially lowered by an amount Δh. No exact solution has been found for this problem. Hence, some simplifications are needed. First, consider steady flow to a point sink as illustrated in Figure 4.2b.

Using spherical coordinates and because of spherical symmetry, the flux directed to the sink q_r at a distance r can be found by dividing the total flow Q by the surface of a sphere with radius r, or

$$q_r = -K\frac{\partial h}{\partial r} = -\frac{Q}{4\pi r^2}, \tag{4.18}$$

which enables to find h as a function of r

$$h = \frac{-Q}{4\pi K r} + C_1. \tag{4.19}$$

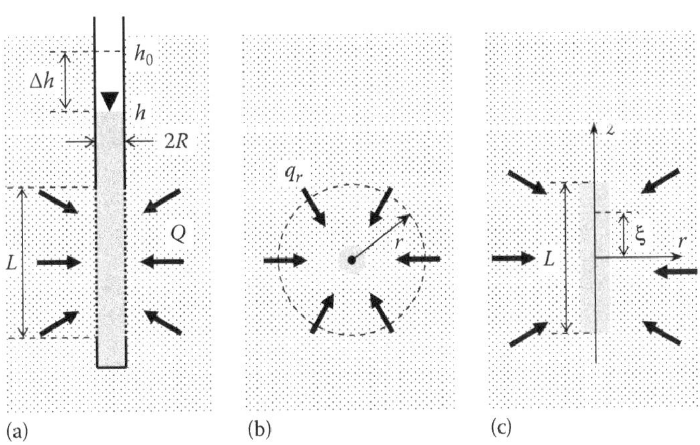

FIGURE 4.2 Groundwater flow through a piezometer screen when the water level in the piezometer is artificially lowered: (a) actual situation, (b) approximation by a point sink, and (c) approximation by a line sink.

The integration constant can be obtained by imposing that far from the sink, $r \to \infty$, the head remains unchanged, i.e., $h = h_0$, yielding

$$h = h_0 - \frac{Q}{4\pi K r}, \tag{4.20}$$

or

$$\Delta h = h_0 - h = \frac{Q}{4\pi K r}. \tag{4.21}$$

Because a point sink is not a good representation of a piezometer screen, we now consider a line source sink, as shown in Figure 4.2c. The line sink has a length L and extracts a total flow Q, to mimic the piezometer screen as well as possible. To find the heads, we use cylindrical coordinates and consider the line sink to be composed of an infinite series of point sinks distributed along z-axis. We use for this an additional coordinate ξ, which runs from $-L/2$ to $+L/2$ as shown in Figure 4.2c. Hence, each point sink extracts an elementary flow $Qd\xi/L$ and the distance from any point to a sink is given by $\sqrt{(z-\xi)^2 + r^2}$. The resulting head distribution can be obtained by summing up all the contributions of each point sink, i.e.

$$\begin{aligned}\Delta h = \int \mathrm{d}h &= \int_{-L/2}^{L/2} \frac{Q}{4\pi K L \sqrt{(z-\xi)^2 + r^2}} \mathrm{d}\xi \\ &= \frac{Q}{4\pi K L}\left[\sinh^{-1}\left(\frac{\frac{1}{2}L + z}{r}\right) + \sinh^{-1}\left(\frac{\frac{1}{2}L - z}{r}\right)\right],\end{aligned} \tag{4.22}$$

where $\sinh^{-1}$ is the inverse hyperbolic sine.

Finally, the head drop in the piezometer is approximated by equating the above formula for $z = 0$ and $r = R$, yielding

$$\Delta h = \frac{Q}{2\pi K L}\sinh^{-1}\left(\frac{L}{2R}\right) = \frac{Q}{2\pi K L}\ln\left[\frac{L}{2R} + \sqrt{1 + \left(\frac{L}{2R}\right)^2}\right], \tag{4.23}$$

which can be approximated for L/R large as

$$\Delta h = \frac{Q}{2\pi K L}\ln\left(\frac{L}{R}\right). \tag{4.24}$$

This equation was presented by Hvorslev (1951) and forms the basis for the so-called piezometer test, or slug test, to determine the hydraulic conductivity in the field with a piezometer. As explained by, for instance, Hvorslev (1951) and Freeze and Cherry (1979), the head difference and the flow rate can be obtained by measuring the change in the water level inside the piezometer and subsequently be used to calculate the hydraulic conductivity.

4.3 THEIS WELL FLOW EQUATION

Consider a fully penetrating well in a homogeneous, horizontal and infinite in radial direction, confined aquifer, as shown in Figure 4.3. Initially, the head is constant and equal to h_0. At time $t=0$, pumping starts with a constant rate Q.

The problem is described by the three-dimensional groundwater flow equation, given by Equation 2.4. Because the problem is axially symmetric, it can best be described in cylindrical coordinates (r,θ,z), as in Equation 2.6, but the term in θ disappears because of the axial symmetry and in z because the flow to the well is strictly horizontal; hence, the flow equation reduces to

$$S_s \frac{\partial h}{\partial t} = \frac{1}{r}\frac{\partial}{\partial r}\left(rK_h \frac{\partial h}{\partial r}\right). \tag{4.25}$$

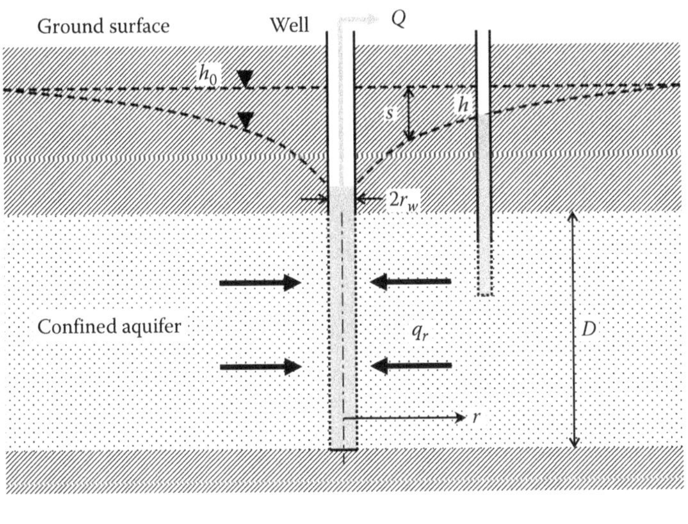

FIGURE 4.3 Groundwater flow in a confined aquifer to a pumped fully penetrating well.

The initial condition is

$$h(r,0) = h_0, \tag{4.26}$$

and the boundary conditions are: at infinity, the given initial head

$$h(\infty,t) = h_0, \tag{4.27}$$

and at the well screen, the given radial flux

$$q_r(r_w,t) = -K_h \left.\frac{\partial h}{\partial r}\right|_{r=r_w} = -\frac{Q}{2\pi r_w D}, \tag{4.28}$$

where r_w is the radius of the well screen. Notice that the term on the right-hand side is negative, because the flow is directed out of the aquifer. However, the boundary condition at the well is rather complicated; hence, this will be simplified by assuming that the radius of the well tends to zero, or

$$\lim_{r\to 0}\left(r\frac{\partial h}{\partial r}\right) = \frac{Q}{2\pi D K_h}. \tag{4.29}$$

The solution is obtained by means of the Boltzmann transformation, by introducing a new independent variable $\lambda = r^2/t$. The transformation formulae are

$$\frac{\partial h}{\partial t} = \frac{\mathrm{d}h}{\mathrm{d}\lambda}\frac{\partial \lambda}{\partial t} = -\frac{r^2}{t^2}\frac{\mathrm{d}h}{\mathrm{d}\lambda} = -\frac{\lambda}{t}\frac{\mathrm{d}h}{\mathrm{d}\lambda}, \tag{4.30}$$

$$r\frac{\partial h}{\partial r} = r\frac{\mathrm{d}h}{\mathrm{d}\lambda}\frac{\partial \lambda}{\partial r} = \frac{2r^2}{t}\frac{\mathrm{d}h}{\mathrm{d}\lambda} = 2\lambda\frac{\mathrm{d}h}{\mathrm{d}\lambda}, \tag{4.31}$$

$$\frac{1}{r}\frac{\partial}{\partial r}\left(r\frac{\partial h}{\partial r}\right) = \frac{1}{r}\frac{\partial}{\partial r}\left(2\lambda\frac{\mathrm{d}h}{\mathrm{d}\lambda}\right) = \frac{1}{r}\frac{\mathrm{d}}{\mathrm{d}\lambda}\left(2\lambda\frac{\mathrm{d}h}{\mathrm{d}\lambda}\right)\frac{\partial \lambda}{\partial r} = \frac{4}{t}\frac{\mathrm{d}}{\mathrm{d}\lambda}\left(\lambda\frac{\mathrm{d}h}{\mathrm{d}\lambda}\right), \tag{4.32}$$

so that Equation 4.25 is transformed in

$$4K_h\frac{\mathrm{d}}{\mathrm{d}\lambda}\left(\lambda\frac{\mathrm{d}h}{\mathrm{d}\lambda}\right) + S_s\left(\lambda\frac{\mathrm{d}h}{\mathrm{d}\lambda}\right) = 0. \tag{4.33}$$

The boundary conditions given by Equations 4.26 and 4.27 are transformed to

$$\lim_{\lambda\to\infty}(h) = h_0, \tag{4.34}$$

and the boundary condition given by Equation 4.29 is transformed to

$$\lim_{\lambda \to 0}\left(\lambda \frac{\mathrm{d}h}{\mathrm{d}\lambda}\right) = \frac{Q}{4\pi D K_h}. \tag{4.35}$$

Equation 4.33 is a first-order differential equation in $\lambda \mathrm{d}h/\mathrm{d}\lambda$ and can be solved as

$$\lambda \frac{\mathrm{d}h}{\mathrm{d}\lambda} = C_1 \exp\left(-\frac{S_s\lambda}{4K_h}\right) \tag{4.36}$$

where C_1 is an integration constant that can be determined with the boundary condition given by Equation 4.35, yielding

$$\lambda \frac{\mathrm{d}h}{\mathrm{d}\lambda} = \frac{Q}{4\pi D K_h} \exp\left(-\frac{S_s\lambda}{4K_h}\right). \tag{4.37}$$

This can be solved for h as

$$h = \frac{Q}{4\pi D K_h} \int \lambda^{-1} \exp\left(-\frac{S_s\lambda}{4K_h}\right) \mathrm{d}\lambda + C_2, \tag{4.38}$$

where C_2 is a second integration constant that can be determined with the boundary condition given by Equation 4.34, yielding

$$h(r,t) = h_0 - \frac{Q}{4\pi D K_h} \int_{r^2/t}^{\infty} \lambda^{-1} \exp\left(-\frac{S_s\lambda}{4K_h}\right) \mathrm{d}\lambda. \tag{4.39}$$

This equation was derived by Theis (1935) through analogy with the flow and storage of heat in solids. It is usually written using the drawdown $s = h_0 - h$ [L] and overall aquifer properties transmissivity, T, and storativity, S, as

$$s = \frac{Q}{4\pi T} W\left(\frac{Sr^2}{4Tt}\right), \tag{4.40}$$

where the Theis well function, $W(u)$, is given by

$$W(u) = \int_{u}^{\infty} \lambda^{-1} e^{-\lambda} \mathrm{d}\lambda. \tag{4.41}$$

The Theis well flow equation is probably the most fundamental equation in groundwater flow and the most commonly used, especially for the design of wells and the analysis of pumping tests.

4.4 HANTUSH LEAKY WELL FLOW EQUATION

Consider a fully penetrating well in a homogeneous and perfectly horizontal semiconfined or leaky aquifer, as shown in Figure 4.4. The setting and conditions are very similar to those of the confined aquifer considered in the previous case, but an essential difference is that the pumping of the well in the semiconfined aquifer will induce leakage through the aquitard. It is assumed that this water is coming from an aquifer on top, which can yield sufficient water without being depleted. Hence, it is assumed that initially the head in both aquifers is constant and equal to h_0, and that even when the well is pumping the head in the top aquifer will remain unchanged.

The leakage through the aquitard will induce vertical flows in the semiconfined aquifer, resulting in a rather complex flow pattern. In order to simplify

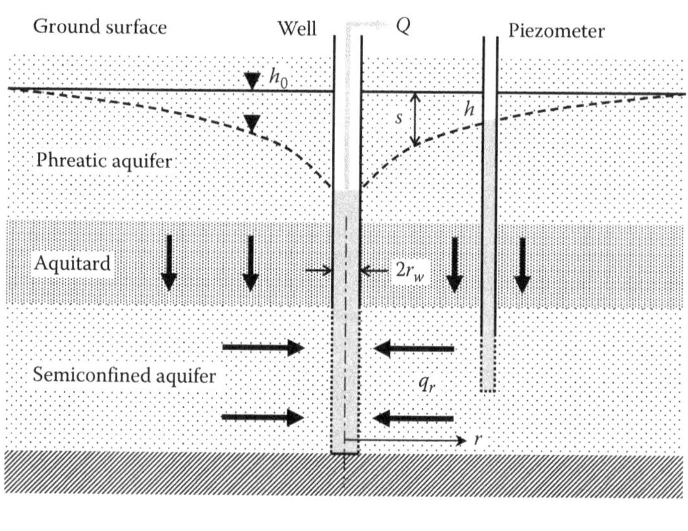

FIGURE 4.4 Groundwater flow in a semiconfined aquifer to a pumped fully penetrating well and induced leakage through the overlying aquitard.

the problem, we will use the hydraulic approach for flow in the semiconfined aquifer, given by Equation 3.12, and for leakage through the aquitard, given by Equation 3.7. Because of axial symmetry, the flow equation becomes

$$S\frac{\partial h}{\partial t} = \frac{1}{r}\frac{\partial}{\partial r}\left(rT\frac{\partial h}{\partial r}\right) + C\left(h_0 - h\right). \tag{4.42}$$

The last term in this equation represents the leakage through the aquitard. The boundary conditions are similar to those in the previous case, i.e., Equations 4.26, 4.27, and 4.28. The flow equation can also be expressed as a function of the drawdown, $s = h_0 - h$, as follows:

$$S\frac{\partial s}{\partial t} = \frac{1}{r}\frac{\partial}{\partial r}\left(rT\frac{\partial s}{\partial r}\right) - Cs = T\frac{\partial^2 s}{\partial r^2} + \frac{T}{r}\frac{\partial s}{\partial r} - Cs, \tag{4.43}$$

with initial condition

$$s\left(r,0\right) = 0, \tag{4.44}$$

and boundary conditions

$$s\left(\infty,t\right) = 0, \tag{4.45}$$

$$\lim_{r \to 0}\left(r\frac{\partial s}{\partial r}\right) = -\frac{Q}{2\pi T}, \tag{4.46}$$

The solution is obtained with the Laplace transform

$$\tilde{s}\left(r,p\right) = \int_0^\infty s\left(r,t\right)e^{-pt}\mathrm{d}t, \tag{4.47}$$

where $\tilde{s}\left(r,p\right)$ is the Laplace transform of $s(r,t)$ and p is the transformation parameter. After transformation, Equation 4.43 becomes

$$\frac{\partial^2 \tilde{s}}{\partial r^2} + \frac{1}{r}\frac{\partial \tilde{s}}{\partial r} - \frac{Sp + C}{T}\tilde{s} = 0, \tag{4.48}$$

where we have made use of the initial condition given by Equation 4.44. Equation 4.48 is the modified Bessel equation of order zero with general solution

$$\tilde{s} = C_1 I_0\left(r\sqrt{\frac{Sp+C}{T}}\right) + C_2 K_0\left(r\sqrt{\frac{Sp+C}{T}}\right), \tag{4.49}$$

where

I_0 is the modified Bessel function of the first kind of order zero
K_0 is the modified Bessel function of the second kind of order zero
C_1 and C_2 are arbitrary constants, which have to be determined from the boundary conditions

The Laplace transform of the boundary condition given by Equation 4.45 is

$$\tilde{s}(\infty, p) = 0, \tag{4.50}$$

which implies that C_1 should be zero, since $I_0(\infty)=\infty$ and $K_0(\infty)=0$. The Laplace transform of the boundary condition given by Equation 4.46 is

$$\lim_{r\to 0}\left(r\frac{\partial \tilde{s}}{\partial r}\right) = -\frac{Q}{2\pi T p}, \tag{4.51}$$

and as the derivative of $K_0(x)$ is $K_1(x)$, the modified Bessel function of the second kind of order one, which tends to x^{-1} for $x \to 0$; the second constant can be determined and substituted in Equation 4.48 yielding

$$\tilde{s} = \frac{Q}{2\pi T p} K_0\left(r\sqrt{\frac{Sp+C}{T}}\right). \tag{4.52}$$

Finally, making use of properties and tables of the Laplace transform (e.g., Abramowitz and Stegun, 1970; Zwillinger, 2002) to back-transform the solution results in

$$s = \frac{Q}{4\pi T}\int_0^t \tau^{-1} \exp\left(-\frac{Sr^2}{4T\tau} - \frac{C\tau}{S}\right) d\tau. \tag{4.53}$$

After substituting $\lambda = Sr^2/4T\tau$ as integration variable, this becomes

$$s = \frac{Q}{4\pi T} W\left(\frac{Sr^2}{4Tt}, \frac{r}{B}\right), \tag{4.54}$$

where $B = \sqrt{T/C}$ is the leakage factor [L] and $W(u, \beta)$ is the Hantush leaky well function, given by

$$W(u,\beta) = \int_u^\infty \lambda^{-1} \exp\left(-\lambda - \beta^2/4\lambda\right) d\lambda. \tag{4.55}$$

This equation was first developed by Hantush and Jacob (1955) and is extensively used for design and analysis of wells in semiconfined aquifers. Although it resembles very much the Theis equation, it cannot be considered as a straightforward extension of the Theis equation, because the Theis equation is an exact solution of the 3-D groundwater flow equation (albeit trivial because there are no vertical flow components), while the Hantush leaky well flow equation is based on the hydraulic flow approximation, where vertical head gradients are ignored. An exact solution of well flow in a semiconfined aquifer is much more difficult but possible to obtain and has been given by Bruggeman (1999, p. 392), which in our notation becomes

$$s = \frac{Q}{2\pi T} \sum_{i=0}^{\infty} \frac{\sin\alpha_i}{\alpha_i + \sin\alpha_i \cos\alpha_i} W\left(\frac{Sr^2}{4Tt}, \frac{\alpha_i r}{D}\right) \cos\left(\frac{\alpha_i z}{D}\right), \tag{4.56}$$

where z is measured from the base of the aquifer and α_i ($i = 0, 1, 2, \ldots$) are the roots of $\alpha \tan\alpha = D^2/B^2$. Comparing Equation 4.56 with Equation 4.54 clearly illustrates the complexity that arises from considering fully 3-D groundwater flow versus the hydraulic approximation.

References

Abramowitz, M. and Stegun, I.A. 1970. *Handbook of Mathematical Functions*. Dover Publications Inc., New York.

Batu, V. 2010. Comment on "Methods to derive the differential equation of the free surface boundary" by Chen, C., Kuang, X. and Jiao, J.J., 2010. *Groundwater* 48, 489–490.

Bear, J. 1972. *Dynamics of Fluids in Porous Media*. American Elsevier Publ. Co., Inc., New York.

Bear, J. 1979. *Hydraulics of Groundwater*. McGraw-Hill Inc., New York.

Bear, J. 1988. *Dynamics of Fluids in Porous Media*. Dover Publ., Inc., New York (reprint from 1972 Edition by American Elsevier).

Bear, J. and Cheng, H.-D. 2010. Comment on "Methods to derive the differential equation of the free surface boundary" by Chen, C., Kuang, X. and Jiao, J.J., 2010. *Groundwater* 48, 486–489.

Bear, J. and Corapcioglu, M.Y. (eds.) 1984. *Fundamentals of Transport Phenomena in Porous Media*. Martinus Nijhoff Publ., Dordrecht, the Netherlands.

Bear, J. and Verruijt, A. 1987. *Modeling Groundwater Flow and Pollution*. D. Reidel Publ. Co., Dordrecht, the Netherlands.

Biot, M.A. 1955. Theory of elasticity and consolidation for a porous anisotropic solid. *J. Appl. Phys.* 26, 182–185.

Borcea, L. 2002. Electrical impedance tomography. *Inverse Probl.* 18, R99–R136.

Bruggeman, G.A. 1999. *Analytical Solutions of Geohydrological Problems*. Elsevier, Amsterdam, the Netherlands.

Calderón, A.P. 1980. On an inverse boundary value problem. In *Seminar on Numerical Analysis and Its Applications to Continuum Physics*, eds. W.H. Meyer and M.A. Raupp, pp. 65–73. Sociedade Brasileira de Matematica, Rio de Janeiro, Brazil.

Charny, I.A. 1951. A rigorous derivation of Dupuit's formula for unconfined seepage with seepage surface (in Russian). *Dokl. Akad. Nauk. SSSR* 79(6), 937–940.

Cooper, H.H. 1966. The equation of groundwater flow in fixed and deforming coordinates. *J. Geophys. Res.* 71, 4785–4790.

Dagan, G. 1989. *Flow and Transport in Porous Formations*. Springer-Verlag, Berlin, Germany.

Dagan, G. and Neuman, S.P. (eds.) 1997. *Subsurface Flow and Transport: A Stochastic Approach*. International Hydrology Series. Cambridge University Press, Cambridge, U.K.

de Marsily, G. 1986. *Quantitative Hydrogeology*. Academic Press, New York.

De Wiest, R.G.M. 1966. On the storage coefficient and the equations of groundwater flow. *J. Geophys. Res.* 71, 1117–1122.

Doherty, J.E. and Hunt, R.J. 2010. Approaches to highly parameterized inversion: A guide to using PEST for groundwater-model calibration. U.S. Geological Survey Scientific Investigations Report 2010-5169, 60pp.

Doherty, J.E., Hunt, R.J., and Tonkin, M.J. 2010. Approaches to highly parameterized inversion: A guide to using PEST for model-parameter and predictive-uncertainty analysis. U.S. Geological Survey Scientific Investigations Report 2010-5211, 71pp.

El-Rawy, M.A., Batelaan, O., and Zijl, W. 2015. Simple hydraulic conductivity estimation by the Kalman filtered double constraint method. *Groundwater* 53(3), 401–413. doi: 10.1111/gwat.12217.

Fetter, C.W. 2001. *Applied Hydrogeology*, 4th edn. Pearson Prentice Hall, Inc., Upper Saddle River, NJ.

Freeze, R.A. and Cherry, J.A. 1979. *Groundwater.* Prentice Hall, Inc., Englewood Cliffs, NJ.

Frind, E.O. 1983. Exact aquitard response functions for multiple aquifer mechanics. In *Flow through Porous Media*, ed. G.F. Pinder, pp. 86–92. CML U.K. Publ., Southampton, U.K.

Gambolati, G. and Freeze, R.A. 1973. Mathematical simulation of the subsidence of Venice: 1. Theory. *Water Resour. Res.* 9, 721–733.

Halek, V. and Svec, J. 1979. *Groundwater Hydraulics.* Elsevier Scientific Publ. Co., Amsterdam, the Netherlands.

Hantush, M.S. 1964. Hydraulics of wells. In *Advances in Hydrosciences*, Vol. 1, ed. V.T. Chow, pp. 281–432. Academic Press, New York.

Hantush, M.S. and Jacob, C.E. 1955. Non-steady radial flow in an infinite leaky aquifer. *Trans. Am. Geophys. Union* 36, 95–100.

Harr, M.E. 1962. *Groundwater and Seepage.* McGraw-Hill Inc., New York (reprinted in 1990 by Dover Publ. Inc., New York).

Hill, M.C. and Tiedeman, C.R. 2007. *Effective Groundwater Model Calibration: With Analysis of Data, Sensitivities, Predictions and Uncertainty.* John Wiley & Sons, Inc., Hoboken, NJ.

Holzbecher, E.K. 1998. *Modeling Density-Driven Flow in Porous Media.* Springer-Verlag, Berlin, Germany.

Huisman, L. 1972. *Groundwater Recovery.* Macmillan Press Ltd., London, U.K.

Hvorslev, M.J. 1951. Time lag and soil permeability in ground-water observations. U.S. Army Corps of Engineers, Waterways Experimental Station, Bulletin 36, 50pp.

Jacob, C.E. 1940. On the flow of water in an elastic artesian aquifer. *Trans. Am. Geophys. Union* 2, 574–586.

Jacob, C.E. 1950. Flow of groundwater. In *Engineering Hydraulics*, ed. H. Rouse, pp. 321–386. John Wiley & Sons, New York.

Jamali, M. and Seymour, B. 2011. Comment on "Methods to derive the differential equation of the free surface boundary" by Chen, C., Kuang, X. and Jiao, J.J., 2010. *Groundwater* 49, 139–141.

Kohn, R.V. and Vogelius, M. 1987. Relaxation of a variational method for impedance computed tomography. *Comm. Pure Appl. Math.* 40(6), 745–777.

Kruseman, G.P., De Ridder, N.A., and Verweij, J.M. 1990. *Analysis and Evaluation of Pumping Test Data.* International Institute for Land Reclamation and Improvement/ILRI, Wageningen, the Netherlands.

McWhorter, P.B. and Sunada, D.K. 1977. *Groundwater Hydrology and Hydraulics*. Water Resources Publ., Colorado, OH.

Mohammed, G.A., Zijl, W., Batelaan, O., and De Smedt, F. 2009. Comparison of two mathematical models for 3D groundwater flow: Block-centered heads and edge-based stream functions. *Transp. Porous Media* 79, 469–485.

Neuman, S.P. 1972. Theory of flow in unconfined aquifers considering delayed response of the water table. *Water Resour. Res.* 8, 1031–1045.

Neuman, S.P. 2011. Comment on "Methods to derive the differential equation of the free surface boundary" by Chen, C., Kuang, X. and Jiao, J.J., 2010. *Groundwater* 49, 135–136.

Olmstead, W.E. 1968. Force relationships and integral representations for the viscous hydrodynamical equations. *Arch. Rational Mech. Anal.* 31, 380–390.

Pinder, G.P. (ed.) 1983. *Flow through Porous Media*. CML Publ., Southampton, U.K.

Polubarinova-Kochina, P.Y. 1962. *Theory of Ground Water Movement*. Princeton Univ. Press, Princeton, NJ.

Rubin, Y. 2003. *Applied Stochastic Hydrogeology*. Oxford University Press, Oxford, U.K.

Strack, O.D.L. 1984. Three-dimensional streamlines in Dupuit-Forchheimer models. *Water Resour. Res.* 20, 812–822.

Strack, O.D.L. 1989. *Groundwater Mechanics*. Prentice Hall, Inc., Englewood Cliffs, NJ.

Strack, O.D.L. and Verruijt, A. 2011. Comment on "Methods to derive the differential equation of the free surface boundary" by Chen, C., Kuang, X. and Jiao, J.J., 2010. *Groundwater* 49, 136–139.

Sun, N.-Z. 2004. *Inverse Problems in Groundwater Modeling*. Kluwer Academic Publ., Dordrecht, the Netherlands.

Theis, C.V. 1935. The relation between the lowering of the piezometric surface and the rate and duration of discharge of a well using groundwater storage. *Trans. Am. Geophys. Union* 16, 519–524.

Todd, D.K. 1980. *Groundwater Hydrology*, 2nd edn. Wiley, New York.

Tóth, J. 1962. A theory of groundwater motion in small drainage basins in Central Alberta. *J. Geophys. Res.* 57(11), 4375–4387.

Tóth, J. 1963. A theoretical analysis of groundwater flow in small drainage basins. *J. Geophys. Res.* 68(16), 4795–4812.

Tóth, J. 2009. *Gravitational Systems of Groundwater Flow*. Cambridge University Press, Cambridge, U.K.

Verruijt, A. 1969. Elastic storage of aquifers. In *Flow through Porous Media*, ed. J.M. De Wiest, pp. 331–376. Academic Press, New York.

Verruijt, A. 1970. *Theory of Groundwater Flow*. Macmillan, London, U.K.

Walton, W.A. 1970. *Groundwater Resource Evaluation*. McGraw-Hill, Inc., New York.

Wang, X.-S. 2011. Comment on "Methods to derive the differential equation of the free surface boundary" by Chen, C., Kuang, X. and Jiao, J.J., 2010. *Groundwater* 49, 133–135.

Zijl, W. and Nawalany, M. 1993. *Natural Groundwater Flow*. Lewis Publishers, Boca Raton, FL.

Zwillinger, D. (ed.) 2002. *CRC Standard Mathematical Tables and Formulae*. CRC Press, Boca Raton, FL.

FOR FURTHER INFORMATION

An outstanding work on the mathematical aspects of groundwater hydrology is Bear (1972). This work gives a comprehensive coverage of the dynamics of fluid flow through porous media. This book has been reprinted (Bear, 1988) and is highly recommended for scientists and engineers. Furthermore, some parts have been revised or updated and republished, such as Bear (1979) dealing with principles of regional groundwater management and with special emphasis on the hydraulic approach to groundwater flow and Bear and Verruijt (1987) discussing mathematical and numerical modeling of groundwater flow and pollution transport. Another classical work is Polubarinova-Kochina (1962), which gives an extensive overview of mathematical aspects of groundwater flow, with many analytical solutions to practical problems. Other notable works are McWhorter and Sunada (1977), Freeze and Cherry (1979), Todd (1980), and de Marsily (1986). A mathematical in-depth discussion of groundwater flow including perturbation calculus and the kinematics and dynamics of the water table is presented in Zijl and Nawalany (1993). Stochastic groundwater flow approach is discussed by Dagan (1989), Dagan and Neuman (1997), and Rubin (2003). A very comprehensive overview of analytical solutions of geohydrological problems and analytical solution methods is given in Bruggeman (1999).

Textbooks on applied hydrogeology form a valuable complement to this chapter and the earlier mentioned books on the laws of groundwater flow. Applied hydrogeology deals with the worldwide wealth of information about the possible hydrogeological settings in which groundwater flow equations and numerical models can be applied in a practical sense; see, for instance, Fetter (2001). In such textbooks, the equations are generally based on the hydraulic approach, i.e., on a division of the subsurface in aquifers and aquitards with their own equations, with recharge flux as the top boundary condition and with hydraulic head as the primary variable. A notable exception is Tóth (2009), who presents the analysis of basin hydrodynamic systems in relation to their contained resources. In this approach, the groundwater flows through the three-dimensional subsurface—or a two-dimensional cross section—using the three- or two-dimensional basic equations (not based on the hydraulic approach), generally with the head specified as the top boundary condition and with focus on groundwater velocity as the primary variable. The basin hydrodynamics approach is not only a way to show the practical relevance of the fully three-dimensional approach, but it brings also together the knowledge and technologies of water geoscience and the energy (oil and gas) geoscience world.

New developments can be found in journals. The well-known ones are *Water Resources Research*, published by the American Geophysical Union, and *Journal of Hydrology of the European Geophysical Society* and *Advances in Water Resources*, both published by Elsevier; these journals cover all scientific hydrology subjects. Other interesting journals are *Hydrogeology Journal of the International Association of Hydrogeologists* and *Groundwater of the U.S. National Ground Water Association*, which cover all aspects of hydrogeology. The most specialized journal in the field of mathematical groundwater hydrology as well as flow and transport through porous media for nonhydrological applications is *Transport in Porous Media*, published by Kluwer Academic Publishers.

Index

For Product Safety Concerns and Information please contact our EU representative GPSR@taylorandfrancis.com
Taylor & Francis Verlag GmbH, Kaufingerstraße 24, 80331 München, Germany

www.ingramcontent.com/pod-product-compliance
Lightning Source LLC
Chambersburg PA
CBHW060740150926
51501CB00003B/50
* 9 7 8 1 1 3 8 5 7 8 8 8 3 *